T0280284

Universitext

Universitext

Editors (North America): J.H. Ewing, F.W. Gehring, and P.R. Halmos

Morton L. Curtis

Abstract Linear Algebra

With Revisions by Paul Place

Springer-Verlag
New York Berlin Heidelberg
London Paris Tokyo Hong Kong

Morton L. Curtis (1921-1989)
Department of Mathematics
Rice University
Houston, TX 77251
USA

AMS Subject Classifications (1980): 15-01, 11R52

Library of Congress Cataloging-in-Publication Data
Curtis, Morton Landers, 1921-
Abstract linear algebra / Morton L. Curtis ; with revisions by Paul Place.
p. cm — (Universitext)
Includes bibliographical references.
ISBN 0-387-97263-3 (alk. paper)
1. Algebras, Linear. I. Place, Paul. II. Title.
QA184.C873 1990
512'.5–dc20 90-33058

Printed on acid-free paper

Photocomposed copy prepared using the author's LaTeX file.
Printed and bound by R.R. Donnelley & Sons, Harrisonburg, Virginia.
Printed in the United States of America.

9 8 7 6 5 4 3 2 1

ISBN 0-387-97263-3 Springer-Verlag New York Berlin Heidelberg
ISBN 3-540-97263-3 Springer-Verlag Berlin Heidelberg New York

Preface

The unexpected and untimely death of the author, and my good friend, Morton Curtis on February 4, 1989 occurred only a short time after the first complete version of this manuscript had been returned by the typist. His intentions with this work were clear: he saw a way of organizing the subject of Linear Algebra around a topic which was not in the general purview and in doing so enhancing both the topic and its audience. Indeed it was characteristic of his approach to mathematics to seek out the elegant ideas, to make them clear and concise, and to teach them with enthusiasm to the widest group possible. It gives me satisfaction to have been able to help insure that this project be completed. That others might gain some enjoyment of mathematics through this book would be a most suitable monument to his memory.

The goal to which this work leads is the Theorem of Hurwitz—that the only normed algebras over the real numbers are the real numbers, the complex numbers, the quaternions, and the octonions. It is, to my knowledge, the only presentation of this material at an "elementary level." It begins from scratch and develops the standard topics of Linear Algebra one would expect from a first course on the subject. It is intended as a text for such a course directed toward students with mathematical leanings; it stresses the complete logical development of the subject while suppressing mention of its many applications to other fields. I think it would also be a valuable reference for mathematicians in general.

Much of the beginning presentation is standard; where options exist, the choice has tended toward the more mathematically sophisticated—particularly where this choice makes a good case for its superiority as, for example, in the treatment of determinants through the introduction of exterior algebras.

The task of preparing the original draft for publication was taken on by my student, Paul Place. In consultation with me, he approached the job of reworking the text to insure clarity, consistency, and completeness with admirable dedication. His efforts include: the inclusion of exercises, rewriting of sections IF, IIE, IIIC, IIID, and large parts of Chapters IV and V, addition of sections IIIE and IVA , and all the other details inherent in such a project. The reader will surely benefit from the care he put into polishing this book.

Our thanks also go to Vivian Choi for her excellent work in TEX-ing all phases of this manuscript.

John Hempel
Houston, Texas
December 1989

Introduction

This book covers most of the basic results in linear algebra, but in some ways it is somewhat more ambitious. The main such feature is in Chapter V in which the Hurwitz Theorem is proved. This result has been known since early in this century, but the proofs have just recently been simplified enough for a textbook at this level. The treatment which gave me the courage to include it is due to Reese Harvey and Blaine Lawson ([3]). Our treatment is just a detailed amplification of their appendix.

Another feature is the introduction of exterior algebras to use in defining determinants. This has been standard for years in more advanced courses, but is not yet standard in beginning linear algebra texts. It greatly simplifies the, usually quite messy, more standard treatments of determinants.

Also, we prove the Cayley-Hamilton Theorem in a general setting. This only necessitates a consideration of polynomials over a not-necessarily commutative ring (namely, the ring of $n \times n$ matrices). Since noncommutative algebras will be a way of life for us in Chapter V, it doesn't hurt to make an acquaintance in Chapter III.

All of this is written as a mathematics course—not an applied mathematics course. It is to be hoped that the reader can read and understand the mathematics without applications. Linear algebra has lots and lots of important and interesting applications to many subjects as well as to other things in mathematics. Try to accept it for now as interesting mathematics.

Contents

Chapter 0

Algebraic Preliminaries

A *field* is an algebraic object consisting of a set with two operations on it, addition and multiplication, which are required to satisfy certain conditions. To avoid a certain amount of repetition in stating these conditions (and thus to also make the conditions easier to remember), we first talk about sets with only one operation.

Definition A *group* G is a set G with one operation $((a, b) \to ab)$ satisfying:

(i) the operation is *associative*

$$a(bc) = (ab)c,$$

(ii) there exists an *identity element* $e \in G$ such that for each $a \in G$

$$ae = ea = a, \qquad \text{and}$$

(iii) for each $a \in G$ there is an *inverse element* a^{-1} such that

$$aa^{-1} = e = a^{-1}a.$$

Note that, because of (ii), the empty set cannot be a group. But given a set with exactly one element (a *singleton*) it has precisely one operation on it, and that operation makes it into a group, a *trivial* or *zero* group.

Now conditions (ii) and (iii) leave open certain possibilities. Namely, there may be more than one identity element, and an element of G may have more than one inverse element. Actually, neither of these can happen.

Proposition 1 *A group G has exactly one identity element and each $a \in G$ has exactly one inverse.*

Proof: Suppose e and f are identity elements in G. Then $fe = e$ since f is an identity element, and $fe = f$ since e is an identity element.

Next suppose that b, c are both inverse elements for a. Then

$$b = eb = (ca)b = c(ab) = ce = c. \quad \blacksquare$$

Examples

(1) The set $\mathbf{Z}$ of integers is a group under addition. The operation is associative, zero is the identity element, and $-a$ is the inverse of a.

(2) $\mathbf{Z}$ is not a group under multiplication. The operation is associative and 1 acts as identity, but only 1 and -1 have multiplicative inverses in $\mathbf{Z}$.

(3) The set $\mathbf{Q}$ of rational numbers is a group under addition.

(4) The set $\mathbf{Q} - \{0\}$ (i.e., all nonzero rationals) is a group under multiplication. (Verify this.)

(5) $\mathbf{R}^+$: The set of all real numbers $x > 0$ forms a group under multiplication. (Verify.)

(6) Let $\mathbf{R}^n$ be the set of all ordered n-tuples

$$x = (x_1, x_2, \ldots, x_n)$$

of real numbers. Given $x = (x_1, \ldots, x_n)$, $y = (y_1, \ldots, x_n)$, define $x + y = (x_1 + y_1, \ldots, x_n + y_n)$. This operation on $\mathbf{R}^n$ is associative, $0 = (0, \ldots, 0)$ acts as identity, and the inverse of $x = (x_1, \ldots, x_n)$ is $-x = (-x_1, \ldots, -x_n)$. This makes $\mathbf{R}^n$ into a group.

We will be primarily interested in this course in groups which are *abelian.*

Definition A group G is *abelian group* if always

$$ab = ba.$$

In the examples above, all groups are abelian.

Now we are ready to define *field.*

Definition A set k with an operation of addition $(a+b)$ and an operation of multiplication (ab) is a *field* if the following conditions are satisfied.

(i) k is an abelian group under $+$ (and the identity element for $+$ is denoted by 0).

(ii) Multiplication distributes over addition

$$a(b+c) = ab + ac.$$

(iii) $k - \{0\}$ is an abelian group under multiplication (the identity element is written as 1).

Note that in a field k, the additive identity 0 is a "multiplicative annihilator." That is, for any $a \in k$ we have $a0 = 0$.
Proof: $a0 = a(0+0)$ (since 0 is the additive identity) $= a0 + a0$ (by (ii)). By (i), $a0$ must have some additive inverse, and adding it to both sides of

$$a0 = a0 + a0$$

gives $0 = a0$. ■

Recall that we could not have an empty group, but could have one with only one element. For a field we need at least two elements, 0 and 1. (These could not be equal since 0 annihilates 1, but 1 times 1 equals 1.)[1] There is a field with just these two elements:

Addition Table

+	0	1
0	0	1
1	1	0

Multiplication Table

×	0	1
0	0	0
1	0	1

Exercise. Prove that these operations cannot be anything else.

Proposition 2 *Let k be a field. If $x, y \in k$ and $x \neq 0$ and $y \neq 0$, then $xy \neq 0$.*

Proof: We will show the contrapositive, that is, if $x \neq 0$ but $xy = 0$, then $y = 0$.

Now $x \neq 0$ implies (by (iii)) that x^{-1} exists (with $x^{-1}x = 1$). Then $y = (x^{-1}x)y = x^{-1}(xy) = x^{-1}0 = 0$. ■

The property described in Proposition 2 is "*no divisors of zero*"; i.e., we cannot have $x \neq 0$ and $y \neq 0$, but $xy = 0$. So a field has no divisors of zero.

But, additively, a field can have "torsion." For example, the two-element field has $1 + 1 = 0$. We can define a three-element field (elements: $0, 1, 2$) by

---

[1]Show that if $1 = 0$ we are reduced to one trivial operation and really have just the zero group.

Addition Table

+	0	1	2
0	0	1	2
1	1	2	0
2	2	0	1

Multiplication Table

×	0	1	2
0	0	0	0
1	0	1	2
2	0	2	1

(Verify that this is a field.) Note that in this field $1+1+1=2+1=0$.

Definition The *characteristic* λ of a field k is the least natural number such that

$$\underbrace{1+1+\ldots+1}_{\lambda}=0.$$

If no such natural number exists, we say k has characteristic *zero* (strange terminology?).

There will be cases in which we will want to have $1+1\neq 0$. So from now on when we say "a field k" we will mean "a field k with $1+1\neq 0$."

Definition If k is a field and s is a subset of k, then s is a *subfield* of k if the operations of k make s into a field. Then k is called an *extension field* of s.

Let $p(x)=a_nx^n+a_{n-1}x^{n-1}+\ldots+a_1x+a_0$ be a polynomial with coefficients a_i in a field k. We say that $r\in k$ is a *root* of $p(x)$ if

$$p(r)=a_nr^n+a_{n-1}r^{n-1}+\ldots+a_1r+a_0=0.$$

The field k is *algebraically closed* if every polynomial with coefficients in k has a root r in k. For example, the field $\mathbf{R}$ is not algebraically closed because $p(x)=x^2+1$ has coefficients in $\mathbf{R}$ but there is no real number r such that $p(r)=r^2+1=0$.

However, the field $\mathbf{C}$ of complex numbers (which we are just now going to introduce) is algebraically closed. This fact is known as "*the fundamental theorem of algebra.*"

The remainder of this introductory chapter is devoted to a classical and very important extension field $\mathbf{C}$ of the field $\mathbf{R}$ of real numbers.

The Field $\mathbf{C}$ on $\mathbf{R}^2$

Example (6) of a group was $\mathbf{R}^n$. In particular, for $n=2$ we have $x=(x_1,x_2)$, $y=(y_1,y_2)$ with

$$x+y=(x_1+y_1,x_2+y_2),$$

identity $0=(0,0)$ and the inverse $-x=(-x_1,-x_2)$ for x. We consider $\mathbf{R}\subseteq\mathbf{R}^2$ by assigning to $r\in\mathbf{R}$ the point $(r,0)\in\mathbf{R}^2$. We easily see that this makes $\mathbf{R}$ into a *subgroup* of $\mathbf{R}^2$. (Write out a formal definition of "subgroup.") Can we define multiplication on $\mathbf{R}^2$ so that $\mathbf{R}^2$ becomes a field with $\mathbf{R} \subseteq \mathbf{R}^2$ as a subfield?

Surely the most obvious try is

$$xy = (x_1y_1, x_2y_2).$$

This operation does distribute over addition

$$\begin{aligned} x(y+z) &= (x_1, x_2)(y_1 + z_1, y_2 + z_2) \\ &= (x_1(y_1 + z_1), x_2(y_2 + z_2)) \\ &= (x_1y_1 + x_1z_1, x_2y_2 + x_2z_2) \\ &= xy + xz. \end{aligned}$$

The operation is easily seen to be associative and commutative. It has an identity $(1, 1)$. But inverses are a problem. Indeed, if this gave a field then (by Proposition 2) we would have no divisors of zero, but

$$(1,0)(0,1) = (0,0) = 0 \quad (\text{the additive zero of } \mathbf{R}^2).$$

Another way of stating the difficulty here is to note that $\mathbf{R}^2 - \{0\}$ is not a group under this multiplication. (Because, for example, $(1, 0) \neq 0$, but if $(1, 0)$ had a multiplicative inverse $x = (x_1, x_2)$ we would have $(1, 0)(x_1, x_2) = (1, 1)$. Whereas $0x_2 = 0 \neq 1$.)

But there is a multiplication on $\mathbf{R}^2$ which makes it into a field. In Section A of Chapter V we will show that the following formula is "forced on us."

$$(*) \qquad (a, b)(c, d) = (ac - bd, ad + bc)$$

In the following set of exercises one shows that $(*)$ does the job.

Exercises

(1) A subset of H of a group G is a *subgroup* of G if the operation on G makes H into a group. Prove that $H \subseteq G$ is a subgroup if and only if

(i) $e \in H$, and

(ii) if $a, b \in H$, then $ab^{-1} \in H$.

(2) Given three distinct objects $\{a, b, c\}$ let G be the set of all *permutations* (= one-to-one maps) of $\{a, b, c\}$, (e.g., $a \to c$, $c \to a$, $b \to b$ is a permutation). Define an operation on G to be iteration (first one permutation and then the other). Show that this makes G into a group. Show that this group is not abelian.

(3) For a positive integer m, let

$$\mathbf{Z}_m = \{0, 1, \ldots, m-1\}.$$

$$\text{Define:} \quad +: \mathbf{Z}_m \times \mathbf{Z}_m \to \mathbf{Z}_m$$
$$\text{mult}: \mathbf{Z}_m \times \mathbf{Z}_m \to \mathbf{Z}_m$$

by taking answers modulo m (e.g., $\mathbf{Z}_6 = \{0, 1, 2, 3, 4, 5\}$ and $(3)(5) = 15$ modulo $6 = 3$). Show that $\mathbf{Z}_m$ has no divisors of zero $\Leftrightarrow$ m is a prime.

(4) Show that $(1, 0)$ acts as identity for $(*)$.

(5) Show that $(*)$ extends the multiplication on $\mathbf{R}$ ($\mathbf{R} = \{(r, 0) | r \in \mathbf{R}\} \subseteq \mathbf{R}^2$).

(6) Show that $(*)$ makes $\mathbf{R}^2$ into a field (the field $\mathbf{C}$ of *complex numbers*).

(7) Write (a, b) as $a + bi$ and treat these as polynomials in i with the condition that $i^2 = -1$. Show this gives $(*)$.

(8) Define the *conjugate* of $\alpha = a + bi \in \mathbf{C}$ to be $\overline{\alpha} = a - bi$. (Note that conjugation is the identity map on $\mathbf{R}$.) Prove that for $\alpha, \beta \in \mathbf{C}$ we have

$$\text{(i)} \quad \overline{(\alpha + \beta)} = \overline{\alpha} + \overline{\beta}, \quad \text{and}$$
$$\text{(ii)} \quad \overline{(\alpha\beta)} = \overline{\alpha}\overline{\beta}.$$

Also show that $\alpha\overline{\alpha}$ is a nonnegative real number, and $\alpha\overline{\alpha} = 0$ exactly when $\alpha = 0$. (The real number $|\alpha| = \sqrt{\alpha\overline{\alpha}}$ is called the *absolute value* of α.) Show that if α is real (i.e., $b = 0$), then $|\alpha|$ is just the usual absolute value of a real number. (The definition of $|r|$ for $r \in \mathbf{R}$ is $\sqrt{r^2}$.)

(9) Let $p(x)$ be a polynomial in the indeterminate x with real coefficients. Show that for any $\alpha \in \mathbf{C}$ we have

$$\overline{p(\alpha)} = p(\overline{\alpha}).$$

Use this to prove that if $p(x)$ (real coefficients) has a complex root α (i.e., $p(\alpha) = 0$), then $\overline{\alpha}$ is also a root.

(10) A polynomial $p(x)$ over a field k is *monic* if the highest power of x has coefficient 1. Let $p(x)$ be monic and let $r \in k$. Show that if $p(x)$ is divided by $x - r$, then the remainder is $p(r)$. (Hint: Write $p(x) = x^n + a_{n-1}x^{n-1} + \ldots + a_1x + a_0$ and do long division by $x - r$.) Show that this remains true if $p(x)$ has coefficients in $\mathbf{R}$ and $r \in \mathbf{C}$.

(11) From (10), we see that if $p(x)$ is monic with real coefficients and $\alpha \in \mathbf{C}$ is a root of $p(x)$, then $x - \alpha$ divides $p(x)$. Now prove that a monic polynomial $p(x)$ with real coefficients can be factored into linear and quadratic factors. (Hint: Let $s_1, \ldots, s_m$ be all of the real roots of $p(x)$. Then

$$p(x) = (x - s_1) \ldots (x - s_m)q(x)$$

where $q(x)$ has real coefficients but no real roots. For each complex root α of $q(x)$ we have factors $(x - \alpha)$ and $(x - \overline{\alpha})$. Show that $(x - \alpha)(x - \overline{\alpha})$ has real coefficients.)

(12) Show that $\{a + b\sqrt{3} \mid a, b \in \mathbf{Q}\}$ is a field (a subfield of the field $\mathbf{R}$), but $\{a + b\sqrt{3} \mid a, b \in \mathbf{Z}\}$ is not a field.

Chapter I

Vector Spaces and Linear Maps

A. Vector Spaces

We introduce the concept of a vector space by using the plane $\mathbf{R}^2$ as a model. $\mathbf{R}^2$ is the set of all ordered pairs (a, b) of real numbers. (The fact that we take *ordered* pairs means, for example, that $(1,2) \neq (2,1)$.) We define an operation $+$ of addition on $\mathbf{R}^2$ by

$$(1) \qquad (a, b) + (c, d) = (a + c, b + d).$$

This operation is associative (since addition of real numbers is associative). The ordered pair $0 = (0,0)$ acts as identity for $+$, and relative to this identity, $(-a, -b)$ is the inverse of (a, b). Thus $\mathbf{R}^2$ becomes a group, and it is abelian since the operation is clearly commutative.

We next define a second operation in which we use a real number and an element of $\mathbf{R}^2$ to produce another element of $\mathbf{R}^2$. Namely, if $r \in \mathbf{R}$ and $(a, b) \in \mathbf{R}^2$ we set

$$(2) \qquad r(a, b) = (ra, rb).$$

Terminology:

We call elements of $\mathbf{R}^2$ *vectors*. We call elements of $\mathbf{R}$ *scalars*. We call the operation (1) *vector addition* and the operation (2) *scalar multiplication*.

Definition A *real vector space* V is a set V of elements (called *vectors*) and two operations:

$$\text{vector} + \text{vector} = \text{vector, and}$$
$$(\text{scalar})(\text{vector}) = \text{vector}.$$

These operations must satisfy the following properties:

(i) V is an abelian group under $+$ (identity $= 0$),

(ii) $1v = v$, and

(iii) the two operations are related by

(a) $r(v + w) = rv + rw$,
(b) $(r + s)v = rv + sv$, and
(c) $(rs)v = r(sv)$,

where v, w are vectors; and r, s are scalars.

It is left as a routine exercise to verify that $\mathbf{R}^2$ with the operations (1) and (2) is a real vector space. Also, what we did for $\mathbf{R}^2$, we can clearly do for $\mathbf{R}^3$ ($=$ all ordered triples of real numbers) and, more generally, for $\mathbf{R}^n$, $n = 1, 2, 3, \ldots$.

But these $\mathbf{R}^n$ spaces are not the only real vector spaces. For example, let $a, b \in \mathbf{R}$ and let V consist of all twice differentiable functions f which satisfy

$$\frac{d^2 f}{dx^2} + a\frac{df}{dx} + bf = 0.$$

We define operations

$$\begin{aligned}(f+g)(x) &= f(x) + g(x), \quad \text{and} \\ (rf)(x) &= r(f(x))\end{aligned}$$

and easily verify that V is a real vector space.

By replacing $\mathbf{R}$ by a field k we get the concept of a *vector space over the field* k.

Examples

(1) The field k is a vector space over itself. The vector addition is just addition in k and scalar multiplication is just multiplication in k.

(2) $k^n = n$-tuples of elements of k. This is a vector space over k just as $\mathbf{R}^n$ is a real vector space.

(3) Let V be the set of all continuous functions from $[0, 1]$ to $\mathbf{R}$. We define

$$\begin{aligned}(f+g)(t) &= f(t) + g(t), \quad \text{and for } r \in \mathbf{R}, \\ (rf)(t) &= rf(t).\end{aligned}$$

Since f, g continuous implies that $f+g$ is continuous and rf is continuous, V becomes a real vector space.

Definition If V is a vector space over k and $r_1, \ldots, r_m \in k$ and $v_1, \ldots, v_m \in V$, then the vector

$$r_1 v_1 + \ldots + r_m v_m$$

is called a *linear combination* of $v_1, \ldots, v_m$.

Definition Let V be a vector space over a field k. A subset W of V is called a *subspace* (or *linear subspace*) if the operations of V make W into a vector space over k.

Proposition 1 *A subset W of a vector space V is a subspace $\Leftrightarrow$ any linear combination of elements of W is in W.*

Proof: $\Rightarrow$ If W is a subspace, it is closed under vector addition and scalar multiplication and hence closed under taking linear combinations.

$\Leftarrow$ This is obvious. ∎

Examples

(1) Let D be all real-valued twice differentiable functions defined on $[0,1]$. Given P, Q in D, let W be all solutions $y(x)$ of the differential equation $y'' + P(x)y' + Q(x)y = 0$. Then W is a subspace of D.

(2) Let $\mathbf{R}^\infty$ be the set of all sequences $\{a_i\}$ of real numbers. We add coordinatewise and multiply by a real number coordinatewise. Let $W \subseteq \mathbf{R}^\infty$ be all sequences $\{a_i\}$ such that $\sum a_i^2$ is a convergent series. Then W is a subspace (proved in §B, Chapter IV).

(3) Let P be the set of all polynomials with real coefficients. Let P_n be those of degree $\leq n$. Then P_n is a subspace of P.

Proposition 2 *If V is a vector space and $\{W_\alpha\}_{\alpha \in A}$ is any collection of subspaces, then*

$$W = \bigcap_{\alpha \in A} W_\alpha$$

is a subspace.

Proof: Exercise. ∎

Definition If V is a vector space and A is any subset of V, then *Span* (A) is the intersection of all subspaces of V which contain A.

By propositions 1 and 2 we see that Span (A) is a subspace of V equal to all finite linear combinations of vectors in A.

Definition $A \subseteq V$ *generates* (or *spans*) V if Span $(A) = V$.

Examples

(1) $A = \{(n,0)|n \in \mathbf{Z}\} \subseteq \mathbf{R}^2$ does not generate $\mathbf{R}^2$.

(2) $A = \{(1,0),(0,1)\} \subseteq \mathbf{R}^2$ does generate $\mathbf{R}^2$.

(3) $A = \{(t,t^2,t^3)|t \in \mathbf{R}\} \subseteq \mathbf{R}^3$ generates $\mathbf{R}^3$.

(1) and (2) should be easy. But, at this stage of our development of this subject, (3) should not be obvious. But we already do know how to prove it and we proceed to (somewhat lengthily) do so.

Let (a,b,c) be an arbitrary element of $\mathbf{R}^3$. We want to show that (a,b,c) is some linear combination of elements of A. We will take

$$\begin{aligned} u &= (1,1,1) \\ v &= (2,4,8) \\ w &= (3,9,27) \end{aligned}$$

as elements of A and we assert that the arbitrary element (a,b,c) of $\mathbf{R}^3$ is a linear combination of these three elements of A. We seek real numbers

$$\alpha, \beta, \gamma$$

such that

$$\alpha u + \beta v + \gamma w = (a,b,c).$$

Now

$$\begin{aligned} \alpha u &= (\alpha,\alpha,\alpha) \\ \beta v &= (2\beta,4\beta,8\beta) \\ \gamma w &= (3\gamma,9\gamma,27\gamma). \end{aligned}$$

So we must have

(1) $\alpha + 2\beta + 3\gamma = a$

(2) $\alpha + 4\beta + 9\gamma = b$

(3) $\alpha + 8\beta + 27\gamma = a.$

You do know how to solve these equations for α, β, γ. For example, solve (1) for α ($\alpha = a - 2\beta - 3\gamma$) and plug this into (2) and (3). This gives

(1′) $a + 2\beta + 6\gamma = b$

(2′) $a + 6\beta + 24\gamma = c.$

Next we solve the first for β and plug that into the second. ($2\beta = b - a - 6\gamma$, and putting that in the second gives

$$a + 3(b - a - 6\gamma) + 24\gamma = c\,.)$$

This last equation is easily solved for γ to give

$$\begin{aligned} 6\gamma &= c - 3b + 2a \quad \text{or} \\ \gamma &= \frac{c - 3b + 2a}{6}. \end{aligned}$$

Thus for any $(a, b, c) \in \mathbf{R}^3$ we can find an appropriate γ. But then we can plug this γ back into $1'$ and $2'$ to get

$$\begin{aligned} (1'') \qquad & a + 2\beta + (c - 3b + 2a) = b \\ (2'') \qquad & a + 6\beta + 4(c - 3b + 2a) = c\,. \end{aligned}$$

From $(1'')$ we have $2\beta = 4b - 3a - c$ and plugging this and the value for γ into (1) will give α.

This example was designed to show that this elementary method of "solve-and-plug-in" is really a powerful method in linear algebra. It will settle questions which are not at all intuitively obvious. Like: the set

$$A = \{(t, t^2, t^3) | t \in \mathbf{R}\} \subseteq \mathbf{R}^3$$

does generate $\mathbf{R}^3$.

However powerful the "solve-and-plug-in" method is, it can be tedious if we keep doing it for every problem. So we will develop general theorems which will give us answers without each time resorting to "solve-and-plug-in."

(The example $A = \{(t, t^2, t^3) | t \in \mathbf{R}\}$ is called the *twisted cubic*, and our "solve-and-plug-in" method is often called the "*method of elimination.*")

Exercises

(1) Show that $\{f : \mathbf{R} \to \mathbf{R} | f \text{ continuous}\}$ with operations

$$\begin{aligned} (f + g)(t) &= f(t) + g(t) \\ (rf)(t) &= rf(t) \end{aligned}$$

is a vector space. Call it V.

(i) Let $t_0 \in \mathbf{R}$ and let

$$W = \{f \in V | f(t_0) = 0\}.$$

Show that W is a subspace of V.

(ii) Let $U = \{f \in V | f(t^2) = (f(t))^2 \text{ for all } t \in \mathbf{R}\}$. Show that U is not a subspace of V.

(iii) Let $X = \{f : \mathbf{R} \to \mathbf{R} | f \text{ differentiable}\}$ and show that X is a subspace of V.

(2) If U, W are subspaces of the vector space V, show that the *sum* of U and W

$$U + W = \{u + w | u \in U, \ w \in W\}$$

is also a subspace of V.

(3) If U and W are subspaces of V, show that $U \cup W$ need not be a subspace. However, if $U \cup W$ is a subspace, show that either $U \subseteq W$ or $W \subseteq U$.

(4) Let $\mathbf{R}^\infty$ be the vector space of all real sequences and let $W \subseteq \mathbf{R}^\infty$ be all sequences with only a finite number of nonzero components. Show that W is a subspace of $\mathbf{R}^\infty$.

(5) Show that the set $\{1, (t-1), (t-1)^2, (t-1)^3\}$ generates P_3 the vector space of polynomials of degree ≤ 3.

(6) Suppose A and B are subsets of the vector space V; show that if $A \subseteq B$, then $\mathrm{Span}(A) \subseteq \mathrm{Span}(B)$.

(7) Consider 2×2 square arrays of real numbers (called *matrices*). We denote the totality of these by

$$M_2(\mathbf{R}) = \left\{ \begin{pmatrix} a & b \\ c & d \end{pmatrix} \middle| \, a, b, c, d \in \mathbf{R} \right\}.$$

We make $M_2(\mathbf{R})$ into a vector space (over $\mathbf{R}$) by defining

$$\begin{pmatrix} a & b \\ c & d \end{pmatrix} + \begin{pmatrix} e & f \\ g & h \end{pmatrix} = \begin{pmatrix} a+e & b+f \\ c+g & d+h \end{pmatrix}$$

and

$$r \begin{pmatrix} a & b \\ c & d \end{pmatrix} = \begin{pmatrix} ra & rb \\ rc & rd \end{pmatrix}.$$

(You will note that this is just the vector space $\mathbf{R}^4$ of ordered quadruples of real numbers. But we will later find good reasons for sometimes writing these ordered quadruples as square arrays.)

$A = \begin{pmatrix} a & b \\ c & d \end{pmatrix} \in M_2(\mathbf{R})$ is *diagonal* if $b = 0 = c$. Show that the set $D \subseteq M_2(\mathbf{R})$ of diagonal matrices is a subspace of $M_2(\mathbf{R})$. Do the same for the set T of *upper triangular* matrices ($c = 0$).

(8) Generalize (8) to 3×3 matrices $M_3(\mathbf{R})$.

(9) $A = \begin{pmatrix} a & b \\ c & d \end{pmatrix} \in M_2(\mathbf{R})$ is *singular* if $ad - bc = 0$; otherwise it is *nonsingular*. Show that the set S of singular matrices in $M_2(\mathbf{R})$ is not a subspace of $M_2(\mathbf{R})$. Show that the set NS of nonsingular matrices in $M_2(\mathbf{R})$ is not a subspace of $M_2(\mathbf{R})$.

B. Linear Maps

For two vector spaces V, W (over the same field k), the kind of function from V to W of most importance to us is a *linear map*.

Definition $\phi: V \to W$ is *linear* if it respects linear combinations; i.e., for any $a, b \in k$ and $x, y \in V$ we have

$$\phi(ax + by) = a\phi(x) + b\phi(y). \tag{3}$$

Exercise. We define the *identity map* $I: V \to V$ by $I(v) = v$ for every $v \in V$. Show that I is linear.

Proposition 3 $\phi: V \to W$ is linear $\Leftrightarrow$

(i) $\phi(x + y) = \phi(x) + \phi(y)$, *and*

(ii) $\phi(ax) = a\phi(x)$.

Proof: $\Rightarrow$ (i) follows from (3) by taking $a = 1 = b$ and (ii) follows from (3) by taking $b = 0$.

$\Leftarrow$ $\phi(ax + by) = \phi(ax) + \phi(by)$ by (i), and this equals $a\phi(x) + b\phi(y)$ by (ii). ∎

Note that if ϕ is linear, then $\phi(0) = 0$ (since $\phi(0) = \phi(0 + 0) = \phi(0) + \phi(0)$).

Examples

(1) Suppose $\phi: \mathbf{R} \to \mathbf{R}$ is linear. Then ϕ is completely determined by knowing $\phi(1)$. (If $\phi(1) = a$, then $\phi(b) = \phi(b1) = b\phi(1) = ba$.)

(2) The same proof shows that a linear map ϕ of $\mathbf{R}$ into any (real) vector space is determined by knowing $\phi(1)$.

(3) Let D be the set of all real-valued C^∞ functions defined on $\mathbf{R}$ (i.e., functions with derivatives of all orders). Then D is a real vector space and we have a map

$$\alpha: D \to D$$

defined by letting

$$\alpha(f) = f', \quad \text{the derivative of } f.$$

Then α is a linear map.

(4) With D as in (3), we define

$$\theta : D \to D$$

by $\theta(f)(x) = \int_0^x f(t)dt$. Then θ is linear.

Definition Let $\phi : V \to W$ be linear. The *image* of ϕ is the set of all $\phi(v)$ for $v \in V$. We denote the image of ϕ by $\phi(V)$ or by im ϕ. The *preimage* of $A \subseteq W$, written $\phi^{-1}(A)$, is the set of all vectors in V which map into A under ϕ.

The *kernel* of ϕ, denoted by ker ϕ, is the set of all $v \in V$ such that $\phi(v) = 0$. The linear map ϕ is *surjective* (or *epic*) if $\phi(V) = W$; it is *injective* (or *monic*) if ker $\phi = 0$. A linear map ϕ which is both surjective and injective is an *isomorphism*. If there is an isomorphism from V to W, we say that they are isomorphic and write $V \cong W$.

Exercise. Show that the linear map ϕ is injective $\Leftrightarrow$ ϕ is *one-to-one* (i.e., if $\phi(v) = \phi(w)$, then $v = w$).

Note that ϕ^{-1} is a function from subsets of W to subsets of V; however, it is not a function from $\phi(V)$ to V in the usual sense unless ϕ is monic (i.e., if ker $\phi \neq \{0\}$ and $w \in$ im ϕ (so there is a $v \in V$ such that $\phi(v) = w$), then $\phi^{-1}(w)$ is not well-defined since it consists of the set $\{v + v' | v' \in \ker \phi\}$).

Examples

(1) $\phi : \mathbf{R} \to \mathbf{R}^2$ defined by $\phi(x) = (x, x)$ is linear and is injective, but not surjective.

(2) $\phi : \mathbf{R}^2 \to \mathbf{R}$ defined by $\phi(x, y) = x + y$ is linear and surjective, but not injective.

(3) Let $\phi : \mathbf{R}^2 \to \mathbf{R}^2$ be linear. Denote $\phi(1,0)$ by (a, c) and $\phi(0,1)$ by (b, d), i.e.,

$$\phi(1,0) = (a,c), \quad \phi(0,1) = (b,d).$$

We assert that ϕ in example (3) is an isomorphism $\Leftrightarrow$ $ad - bc \neq 0$. We can prove this by our "solve-and-plug-in" method discussed in Section A. This is a bit tedious, but we will do it in hopes of motivating our development of the general theory of matrices and determinants.

We have $\phi : \mathbf{R}^2 \to \mathbf{R}^2$ linear, and

$$\phi(1,0) = (a,c), \quad \phi(0,1) = (b,d).$$

Suppose $ad - bc = 0$. We show that ϕ is not injective (and hence can't be an isomorphism).

If both a and c are zero, then $\phi(1,0) = (0,0)$ and we are done. If not, then one, say a, is nonzero. Then $(-\frac{b}{a}, 1) \neq (0,0)$, but we claim that $\phi(-\frac{b}{a}, 1) = (0,0)$. Well, $(-\frac{b}{a}, 1) = -\frac{b}{a}(1,0) + (0,1)$, so $\phi(-\frac{b}{a}, 1) = -\frac{b}{a}\phi(1,0) + \phi(0,1) = -\frac{b}{a}(a,c) + (b,d) = (-b, -\frac{bc}{a}) + (b,d) = (0, \frac{-bc+ad}{a}) = (0,0)$. A similar argument works with b, d replacing a, c. Thus we have proved ϕ an isomorphism $\Rightarrow ad - bc \neq 0$ (by proving $ad - bc = 0 \Rightarrow \phi$ not an isomorphism).

Next we need to show $ad - bc \neq 0 \Rightarrow$ (i) ϕ is injective and (ii) ϕ is surjective.

To prove (i), we suppose

$$\phi(x,y) = (0,0) \tag{4}$$

and prove that $(x,y) = (0,0)$. As before, we have $\phi(1,0) = (a,c)$ and $\phi(0,1) = (b,d)$. Then (4) gives

$$\begin{aligned} (1) \qquad ax + by &= 0, \text{ and} \\ (2) \qquad cx + dy &= 0. \end{aligned}$$

Since $ad - bc \neq 0$ one of a, b must be nonzero. Suppose $a \neq 0$. We solve (1) for x and plug into (2) to get

$$y\left(-\frac{cb}{a} + d\right) = 0 \quad \text{or} \quad y\left(\frac{-cb+ad}{a}\right) = 0.$$

Since $\mathbf{R}$ has no divisors of 0 and $\frac{-cb+ad}{a} \neq 0$, we conclude that $y = 0$. Similarly, b and d can't both be zero so we can solve (1) or (2) for y and plug into the other to get $x = 0$. Thus ϕ is injective.

The proof that ϕ is surjective goes in a similar way. Given $(z,w) \in \mathbf{R}^2$ we want (x,y) such that $\phi(x,y) = (z,w)$. We can solve $ax + by = z$ for either x or y and plug into $cx + dy = w$, solve and plug the answer back into $ax + by = z$.

Proposition 4 *If $\phi : V \to W$ is a linear map which is an isomorphism, then ϕ^{-1} is also linear (and thus is also an isomorphism).*

Proof: Let $a, b \in k$ and $x, y \in W$. Since ϕ is one-to-one, there exist unique elements x', y' of V such that $\phi(x') = x, \phi(y') = y$. Then $\phi(ax' + by') = ax + by$, so that

$$\phi^{-1}(ax + by) = ax' + by' = a\phi^{-1}(x) + b\phi^{-1}(y),$$

proving linearity of ϕ^{-1}. ∎

We conclude this section with a brief discussion of the *Algebra of Endomorphisms* of a vector space.

The Algebra End(V)

Definition An *algebra* is a vector space W which also has a multiplication

$$\begin{aligned} W \times W &\to W \\ (v, w) &\to vw \end{aligned}$$

which distributes over vector addition

$$\begin{aligned} u(v+w) &= uv + uw, \quad \text{and} \\ (v+w)u &= vu + wu \end{aligned}$$

and satisfies $v(aw) = a(vw) = (av)w$ for $a \in k$. The algebra has a *unit element* if there is an element $1 \in W$ which acts as a left and right multiplicative identity. It is *associative* if its multiplication is associative.
Examples

(1) $\mathbf{R}$ is an algebra with unit element.

(2) $\mathbf{C}$ is an algebra with unit element.

Indeed, any field is an algebra with unit element, but an algebra with unit element need not be a field because some nonzero elements in an algebra may not have multiplicative inverses. This is true of the algebra

$$\mathrm{End}(V)$$

which we are about to define. This algebra End(V) plays a central role in the study of linear algebra.

Definition A linear map of a vector space V into itself is called an *endomorphism* (or *operator*) of V. We denote by End(V) the set of all endomorphisms of V.

We first define operations on End(V) to make it into a vector space over k (V being a vector space over the field k). Let $\phi, \psi \in \mathrm{End}(V)$. We define $\phi + \psi$ by

$$(+) \quad (\phi + \psi)(v) = \phi(v) + \psi(v) \quad \text{for all } v \in V.$$

For $a \in k$ and $\phi \in \mathrm{End}(V)$ we define $a\phi$ by

$$(\cdot) \quad (a\phi)(v) = a(\phi(v)) \quad \text{for all } v \in V.$$

The operation (+) is easily seen to be associative and commutative. The zero element of End(V) is the endomorphism sending all $v \in V$ to the zero element in V. Relative to this zero in End(V), the additive inverse of $\phi \in \mathrm{End}(V)$ is $-\phi$ defined by $(-\phi)(v) = -\phi(v)$. Clearly $\phi + (-\phi)$ is the zero endomorphism. The other properties of $(+)$ and $(\cdot)$ are routinely verified so that $\text{End}(V)$ is indeed a vector space over k.

Next, we want to make $\text{End}(V)$ into an algebra. For $\phi, \psi \in \text{End}(V)$ their product $\phi\psi$ is defined by $\phi\psi(v) = \phi(\psi(v))$. That is, we just take their composition

$$\begin{gathered} V \xrightarrow{\psi} V \xrightarrow{\phi} V \\ \underbrace{v \longrightarrow \psi(v) \longrightarrow \phi(\psi(v))}_{\phi\psi}. \end{gathered}$$

(We will use the notation $\phi\psi$ and $\phi \circ \psi$ interchangeably.) $\phi\psi$ is linear because

$$\begin{aligned} (\phi\psi)(av + bw) &= \phi(a\psi(v) + b\psi(w)) \text{ since } \psi \text{ is linear} \\ &= a\phi(\psi(v)) + b\phi(\psi(w)) \text{ since } \phi \text{ is linear.} \end{aligned}$$

Also our multiplication distributes over addition. For let $\theta, \phi, \psi \in \text{End}(V)$ and $v \in V$. Then

$$\begin{aligned} \{\theta(\phi + \psi)\}(v) &= \theta((\phi + \psi)(v)) \\ &= \theta(\phi(u) + \psi(v)) = \theta\phi(v) + \theta\psi(v) \\ &= (\theta\phi + \theta\psi)(v). \end{aligned}$$

Thus $\text{End}(V)$ is an algebra. This algebra does have a unit element, i.e., the identity map of V to V. We denote this map by 1,

$$1 : V \to V,$$

and we clearly have $\phi 1 = 1\phi = \phi$ for all $\phi \in \text{End}(V)$.

It is convenient to use ϕ^2 to denote $\phi\phi$, etc. Thus $p(\phi)$ makes sense if $p(x)$ is any polynomial with coefficients from k.

Definition Let A be an algebra with unit element 1. Then any $a \in A$ is called a *unit* if there exists $b \in A$ such that $ab = 1$ and $ba = 1$. (That is, $a \in A$ is a unit if it has a multiplicative inverse b.) Clearly then b is also a unit; furthermore, this b is unique. (Because if also $ac = ca = 1$, then

$$b = 1b = (ca)b = c(ab) = c1 = c.)$$

Proposition 5 *Let A be an associative algebra with unit element 1. Then the set U of all units in A is a group under multiplication.*

Proof: By hypothesis, multiplication is associative, 1 acts as identity and, by definition, each $u \in U$ has a multiplicative inverse. ∎

Since our multiplication in $\text{End}(V)$ is associative (composition is associative by its very definition) we have a group of units in $\text{End}(V)$. We denote it by $GL(V)$ and call it the *general linear group* over V.

Proposition 6 *$\phi \in End(V)$ is a unit $\Leftrightarrow$ ϕ is an isomorphism.*

Proof: $\Rightarrow$ ϕ is a unit so there exists $\psi \in \text{End}(V)$ so that $\phi\psi = 1$ and $\psi\phi = 1$. Now $\phi\psi = 1$ implies ϕ is surjective (and ψ is injective) and $\psi\phi = 1$ implies ϕ is injective (and ψ is surjective) (see exercises).

$\Leftarrow$ If ϕ is an isomorphism, take $\psi = \phi^{-1}$ to see that ϕ is a unit. ∎

Thus, $GL(V)$ consists of those endomorphisms of V which are actually isomorphisms of V.

Exercises

(1) Let $V \xrightarrow{\phi} W$ be linear. Show that $\ker \phi$ is a subspace of V and $\phi(V)$ is a subspace of W.

(2) Give specific linear maps $\phi, \psi : \mathbf{R}^2 \to \mathbf{R}^2$ such that $\phi\psi \neq \psi\phi$.

(3) Let $V \xrightarrow{\phi} W \xrightarrow{\psi} V$ be linear maps such that $V \xrightarrow{\psi\phi} V$ is an isomorphism. Show that ϕ is injective and ψ is surjective.

(4) A linear map $V \xrightarrow{\rho} V$ is *idempotent* if $\rho\rho = \rho$. Show that if ρ is idempotent then ρ acts as the identity on $\rho(V)$. (Such linear maps are called *projections*: ρ projects V onto $\rho(V)$.)

(5) Show that $\phi : \mathbf{R}^2 \to \mathbf{R}^2$ given by $\phi(x, y) = (x, x + y)$ is linear and so is $\psi : \mathbf{R}^3 \to \mathbf{R}$ given by $\psi(x, y, z) = z$. Show that $\theta : \mathbf{R}^2 \to \mathbf{R}$ given by $\theta(x, y) = xy$ is not linear.

(6) Decide whether or not $\phi : \mathbf{R}^2 \to \mathbf{R}^2$ given by $\phi(x, y) = (x + y, 2x - y)$ is an isomorphism. If it is, find a formula for $\phi^{-1}(u, v)$ and show that $\phi \circ \phi^{-1}(u, v) = (u, v)$ and $\phi^{-1} \circ \phi(x, y) = (x, y)$.

(7) If $\phi : V \to W$ is linear, show that $\phi(-v) = -\phi(v)$ for any $v \in V$.

(8) If $\phi : \mathbf{R}^2 \to \mathbf{R}^2$ is defined as in exercise (6), what is $p(\phi)$ where $p(x) = x^2 - 2x + 1$?

C. Bases, Dimension

Let V be a vector space over a field k. For a (finite) linear combination of vectors

$$a_1 v_1 + a_2 v_2 + \ldots + a_m v_m \quad (a_i \in k, v_i \in V)$$

we make a *convention*: namely, if $i \neq j$ then $v_i \neq v_j$. That is, vectors with different subscripts must be different vectors. We can always put a finite linear combination in this form — if a given vector occurs two or more times, we sum up its coefficients and then write it just once with this sum as coefficient.

Definition Let S be a subset of V. The set S is *linearly independent* if no two distinct finite linear combinations of elements of S can be equal vectors.

For example, if $S = \{u, v, w\}$ is linearly independent and

$$au + bv + cw = a'u + b'v + c'w,$$

then it must be that $a = a'$, $b = b'$, $c = c'$. That is, if we don't have $a = a'$, $b = b'$, and $c = c'$, then

$$au + bv + cw \quad \text{and} \quad a'u + b'v + c'w$$

must be different vectors in V.

Lemma *If S is linearly independent and*

$$a_1v_1 + \ldots + a_mv_m = 0 \quad (v_i\text{'s in } S),$$

then $a_1 = a_2 = \ldots = a_m = 0$.
Proof: We have

$$a_1v_1 + \ldots + a_mv_m = 0 = 0v_1 + \ldots + 0v_m. \quad ■$$

Definition If $S \subseteq V$ is not linearly independent, then it is said to be *linearly dependent.*

Proposition 7

(i) *If $0 \in S$, then S is linearly dependent.*

(ii) *If S is linearly dependent and $S \subseteq T \subseteq V$, then T is linearly dependent.*

(iii) *If S is linearly independent and $U \subseteq S \subseteq V$, then U is linearly independent.*

Proof:

(i) $1 \cdot 0 = 0 \cdot 0$ (two distinct, but equal, linear combinations).

(ii) We have $a_1v_1 + \ldots + a_mv_m = 0$, with $v_i \in S$ and not all a_i being zero. Clearly we have the same relation in T since all $v_i \in T$.

(iii) This is the contrapositive of (ii). ∎

Suppose $S = \{v_1, \dots, v_p\}$ is a finite set of vectors in V. How can we proceed to determine if S is linearly dependent or independent?

If $v_1 = 0$, we are done (S is linearly dependent). If $v_1 \neq 0$, then its span is a nonzero subspace of V.

$$\mathrm{Span}(v_1) = \{tv_1 | t \in k\}.$$

If any of $v_2, \dots, v_p$ lie in this span (e.g., $v_j = t_0 v_1$), then $\{v_1, v_j\}$ is linearly dependent, and hence so is S. If none of $v_2, \dots, v_p$ are in $\mathrm{Span}(v_1)$, consider $\mathrm{Span}(v_1, v_2)$. Show that if one of $v_3, \dots, v_p$ is in $\mathrm{Span}(v_1, v_2)$, then S is linearly dependent. If not, look at

$$\mathrm{Span}(v_1, v_2, v_3).$$

If any of $v_4, \dots, v_p$ lie in this span, we have that S is linearly independent, etc. If v_p is not in $\mathrm{Span}(v_1, \dots, v_{p-1})$, we have that S is linearly independent.

Definition A set T is *infinite* if T contains a subset U which can be put into a one-to-one correspondence with $N = \{1, 2, 3, \dots\}$. (Of course, we may have $U = T$.) If T is not infinite it is *finite*.

Definition A vector space V is *infinite-dimensional* if it contains an infinite linearly independent subset. Otherwise V is *finite-dimensional*. Equivalently, we can define V to be finite-dimensional if it is spanned by some finite subset.

Examples

(1) $\mathbf{R}^\infty$ = all sequences of real numbers (Example (2) of IA after Proposition 1). This vector space is infinite-dimensional. For let

$$S = \{(1,0,0,\dots), (0,1,0,0,\dots), (0,0,1,0,\dots), \dots\}.$$

Then S is infinite (let $(1,0,0,\dots) \leftrightarrow 1, (0,1,0,\dots) \leftrightarrow 2$, $(0,0,1,0,\dots) \leftrightarrow 3$, etc.) and linearly independent.

(2) $\mathbf{R}^2$ is finite-dimensional. Indeed, any three vectors in $\mathbf{R}^2$ form a linearly dependent set. (See (i) of Theorem 1 later in this section.)

Definition A set S in a vector space V is a *basis* for V if:

$\mathrm{Span}(S) = V$, and
S is linearly independent.

Example

The set $\{e_1 = (1,0,\ldots,0), e_2 = (0,1,0,\ldots,0), \ldots, e_n = (0,0,\ldots,1)\}$ is a basis for $\mathbf{R}^n$ or, more generally, for k^n (called the *standard basis* for k^n).

Proposition 8 *$S \subseteq V$ is a basis for $V \Leftrightarrow$ Each element of V is a unique linear combination of elements of S.*

Proof: $\Rightarrow$ Since S spans V, any $v \in V$ is a linear combination of elements of S. Since S is linearly independent, this linear combination must be unique.

$\Leftarrow$ The hypothesis implies both spanning and linear independence. ∎

Proposition 9 *Let $\{v_1, \ldots, v_n\}$ be a basis for V, (defined over a field k). If W is a vector space over k and $w_1, \ldots, w_n$ are any n vectors in W, then there exists exactly one linear map*

$$\phi : V \to W$$

such that $\phi(v_i) = w_i$ for $i = 1, \ldots, n$.

Proof: $v \in V$ may be uniquely written as

$$v = a_1 v_1 + \ldots + a_n v_n,$$

so if ϕ is to be linear we must have

$$\phi(v) = a_1 \phi(v_1) + \ldots + a_n \phi(v_n) = a_1 w_1 + \ldots + a_n w_n. \tag{5}$$

Conversely, if ϕ is defined by (5), it is linear and $\phi(v_i) = w_i$. ∎

Proposition 10 *If $A = \{v_1, \ldots, v_m\}$ spans V, then some subset of A is a basis for V.*

Proof: Let $\mathcal{A}$ be the set of all linearly independent sets in A. Let n be the maximal number of elements in any set in $\mathcal{A}$. By changing notation, we may assume that: $\{v_1, \ldots, v_n\}$ is linearly independent, but if $n < i \leq m$, then $\{v_1, \ldots, v_n, v_i\}$ is linearly dependent. Thus we have a relation

$$a_1 v_1 + \ldots + a_n v_n + a_i v_i = 0$$

with not all coefficients zero. If $a_i = 0$, we contradict the linear independence of $\{v_1, \ldots, v_n\}$. So we can solve for v_i ($v_i = -\frac{a_1}{a_i} v_1 - \ldots - \frac{a_n}{a_i} v_n$) to show that $v_i \in \text{Span}(v_1, \ldots, v_n)$. Thus $v_{n+1}, \ldots, v_m$ are all in $\text{Span}(v_1, \ldots, v_n)$ so $\{v_1, \ldots, v_n\}$ is a basis for V. ∎

Theorem 1 *Let $\{v_1, \ldots, v_n\}$ be a basis for V. Then*

(i) *any $w_1, \ldots, w_n$, w_{n+1} in V are linearly dependent, and*

(ii) *any $u_1, \ldots, u_{n-1}$ fail to span V.*

Proof: (For (i) we really just need that $\{v_1, \ldots, v_n\}$ spans V.) We have

$$\begin{aligned} w_1 &= a_{11}v_1 + \ldots + a_{1n}v_n \\ &\vdots \\ w_{n+1} &= a_{n+1,1}v_1 + \ldots + a_{n+1,n}v_n. \end{aligned}$$

If $w_1 = 0$, we are done. If $w_1 \neq 0$, some a_{1j} must be nonzero and we can solve the first equation for v_j (in terms of the other v's and w_1) and plug this into the remaining n equations.

The second equation has w_1, w_2 and all but one of the v's in it. If $w_2 = 0$ or if w_1 and w_2 have the only nonzero coefficients, then $\{w_1, w_2\}$ is linearly dependent. If not we can solve for some v and plug this into the remaining $n-1$ equations.

By now the pattern of the argument should be clear. Either $\{w_1, w_2, w_3\}$ is linearly dependent or we can solve for some v and plug that into the remaining equations. After n steps we run out of v's and have a nontrivial linear dependence relation among $w_1, \ldots, w_{n+1}$.

The argument for (ii) is similar. If $u_1, \ldots, u_{n-1}$ did span V we would have

$$\begin{aligned} v_1 &= c_{11}u_1 + \ldots + c_{1,n-1}u_{n-1} \\ &\vdots \\ v_n &= c_{n1}u_1 + \ldots + c_{n,n-1}u_{n-1}. \end{aligned}$$

Now $v_1 \neq 0$, so we can solve the first equation for some u and plug in the other equations. We wind up with a nontrivial linear dependence relation among $v_1, \ldots, v_n$, contradicting the hypothesis. ∎

From Proposition 10 and Theorem 1, we see that any finite-dimensional vector space V has a finite basis and that any two bases for V must have the same number of elements. In fact, any infinite-dimensional vector space V also has a basis (called a *Hamel basis*) and any two such bases for V can be put into one-to-one correspondence; we will show existence of bases after we have discussed Zorn's Lemma in section I F.

Definition The number of elements in a basis for the finite-dimensional vector space V is the *dimension* of V, written $\dim V$.

Proposition 11 *Let W be a subspace of the finite-dimensional vector space V. Then any basis for W can be extended to a basis for V.*

Proof: Let $\{w_1, \ldots, w_\rho\}$ be a basis for W. (This must be finite or we would have an infinite linearly independent set in V.) Then $\{w_1, \ldots, w_\rho\}$ is linearly independent in V. If $\text{Span}(w_1, \ldots, w_\rho) = V$, we are done (since $\text{Span}(w_1, \ldots, w_\rho) = W$ so $V = W$).

If not, we take $v_1, \ldots, v_k$ in $V - W$ so that

$$\text{Span}(w_1, \ldots, w_\rho, v_1, \ldots, v_k) = V.$$

We consider linearly independent subsets of $\{w_1, \ldots, w_\rho,\ v_1, \ldots, v_k\}$ and proceed, as in Proposition 10, to get a basis for V containing $\{w_1, \ldots, w_\rho\}$. ∎

This clearly implies $\dim W \leq \dim V$.

Now suppose V and W are finite-dimensional vector spaces over a field k, and

$$\phi : V \to W$$

is a linear map. We have seen that the *kernel* of ϕ

$$\ker \phi = \{v \in V | \phi(v) = 0\}$$

is a subspace of V and the *image* of ϕ

$$\text{im } \phi = \{\phi(v) | v \in V\}$$

is a subspace of W. We now prove an important relation involving dimensions.

Theorem 2 (The Rank Theorem)

$$\dim V = \dim(\ker \phi) + \dim(\text{im } \phi).$$

Proof: Let $\{v_1, \ldots, v_r\} \subseteq V$ be a basis for $\ker \phi$, and $\{w_1, \ldots, w_s\} \subseteq W$ be a basis for $\text{im } \phi$. Choose $u_1, \ldots, u_s \in V$ such that $\phi(u_i) = w_i$ (we can do this by the definition of $\text{im } \phi = \phi(V)$). We claim that $\{v_1, \ldots, v_r,\ u_1, \ldots, u_s\}$ is a basis for V. This will prove the theorem.

First we will show $\{v_1, \ldots, v_r,\ u_1, \ldots, u_s\}$ is linearly independent. Suppose $a_1v_1 + \cdots + a_rv_r + b_1u_1 + \cdots + b_su_s = 0$. Since $v_1, \ldots, v_r$ are in $\ker \phi$, this implies

$$\begin{aligned} \phi(a_1v_1 + \ldots + a_rv_r + b_1u_1 + \ldots + b_su_s) &= \phi(0) = 0 \\ &\| \\ \phi(b_1u_1 + \ldots + b_su_s) & \\ &\| \\ b_1\phi(u_1) + \ldots + b_s\phi(u_s) & \\ &\| \\ b_1w_1 + \ldots + b_sw_s. & \end{aligned}$$

Since $w_1, \ldots, w_s$ are linearly independent in W,

$$b_1 w_1 + \ldots + b_s w_s = 0$$

implies $b_1, \ldots, b_s$ are all zero. But then

$$a_1 v_1 + \ldots + a_r v_r = 0$$

so $a_1, \ldots, a_r$ are all zero. This proves

$$\{v_1, \ldots, v_r,\ u_1, \ldots, u_s\}$$

is linearly independent.

It remains to prove this set of vectors spans V. Let $x \in V$. Then $\phi(x) \in \phi(V)$ so

$$\phi(x) = t_1 w_1 + \ldots + t_s w_s.$$

Set $u = t_1 u_1 + \ldots + t_s u_s$. Since $\phi(u_i) = w_i$, we see that $\phi(x - u) = 0$. Thus $x - u$ is in ker ϕ and is a unique linear combination of $v_1, \ldots, v_r$

$$x - u = m_1 v_1 + \ldots + m_r v_r\,, \quad \text{so}$$
$$x = m_1 v_1 + \ldots + m_r v_r + t_1 u_1 + \ldots + t_s u_s. \quad \blacksquare$$

The rank theorem has two important corollaries.

Corollary 1 *Let V be a finite-dimensional vector space and $\phi : V \to V$ be a linear map. If ϕ is monic, then ϕ is an isomorphism. If ϕ is epic, then ϕ is an isomorphism.*

Proof: If ϕ is monic, the rank theorem implies that $\dim \phi(V) = \dim V$, so $\phi(V) = V$.

If ϕ is epic, $\phi(V) = V$, and the rank theorem implies that $\dim(\ker \phi) = 0$. Thus ϕ is monic. $\blacksquare$

Corollary 2 *Let V and W be finite-dimensional vector spaces of the same dimension. Then $V \cong W$.*

Proof: Let $\{v_1, \ldots, v_n\}$ be a basis for V and let $\{w_1, \ldots, w_n\}$ be a basis for W. By Proposition 9, there is a linear map $\phi : V \to W$ such that $\phi(v_i) = w_i$ for $i = 1, \ldots, n$. Note that ϕ is epic since $\{w_1, \ldots, w_n\}$ spans W.

Now the rank theorem implies that $\ker \phi = \{0\}$, so ϕ is also monic. Hence ϕ is an isomorphism. $\blacksquare$

Corollary 2 says that any n-dimensional vector space V over the field k is essentially k^n since we get an isomorphism

$$V \cong k^n$$

once we choose a basis for V (we already have the standard basis for k^n).

Exercises

(1) Check that

$$\sigma : \mathbf{N} \to \mathbf{N}$$

defined by $\sigma(n) = 2n$ gives a one-to-one correspondence between the set of all natural numbers and the set of all even natural numbers. Use this to prove that:

If a set S is infinite, then it can be put in a one-to-one correspondence with a proper subset of itself.

(2) Determine whether or not $\{(1,1,0),(2,0,-1),(-3,1,1)\}$ is a basis for $\mathbf{R}^3$.

(3) Let $\phi : V \to W$ be linear. Suppose that $v_1, \ldots, v_\rho \in V$ are such that $\phi(v_1), \ldots, \phi(v_\rho)$ are linearly independent in W. Show that $v_1, \ldots, v_\rho$ are linearly independent.

(4) Let $\phi \in \mathrm{End}(V)$ for a finite-dimensional vector space V. Prove that

$$\phi \text{ is monic } \Leftrightarrow \phi \text{ is epic } \Leftrightarrow \phi \text{ is an isomorphism.}$$

(5) Show that $P_n = \{$polynomials with real coefficients of degree $\leq n\}$ is an $(n+1)$-dimensional subspace of the infinite-dimen sional vector space of all real polynomials.

(6) Let V be a vector space over a field k and let U, W be finite-dimensional subspaces of V. Prove that both $U + W = \{u + w | u \in U, w \in W\}$ and $U \cap W$ are finite-dimensional subspaces of V, and

$$\dim(U+W) + \dim(U \cap W) = \dim U + \dim W\ .$$

(Hint: We can find bases:

$$\begin{array}{ll} \{v_1, \ldots, v_p\} & \text{for } U \cap W \\ \{v_1, \ldots, v_p,\ u_1, \ldots, u_q\} & \text{for } U \\ \{v_1, \ldots, v_p,\ w_1, \ldots, w_r\} & \text{for } W\ . \end{array}$$

Show that $\{v_1, \ldots, v_p,\ u_1, \ldots, u_q,\ w_1, \ldots, w_r\}$ is a basis for $U + W$.)

(7) Show that the set of real numbers $\mathbf{R}$ is a vector space over the rational numbers $\mathbf{Q}$. Show that $\mathbf{R}$ is not finite-dimensional as a vector space over $\mathbf{Q}$.

(8) Show that the set of complex numbers $\mathbf{C}$ is a two-dimensional vector space over $\mathbf{R}$.

(9) Define two subspaces U, V of a vector space W to be *complementary* if (i) $U \cap V = \{0\}$, and (ii) $U + V = W$. Prove that U, V are complementary $\iff$ each $w \in W$ may be uniquely represented as $w = u + v$, $u \in U$, $v \in V$.

(10) Let $\{e_1, e_2, e_3\}$ be the standard basis for $\mathbf{R}^3$ and defiine $\phi : \mathbf{R}^3 \to \mathbf{R}$ by $\phi(e_1) = 1$, $\phi(e_2) = 2$, $\phi(e_3) = -1$. Determine the subspaces ker ϕ and im ϕ, and verify the rank theorem in this case.

(11) $\phi : V \to V$ is *nilpotent of order 2* if $\phi\phi$ is the zero endomorphism. Now composition of two such endomorphisms need not be nilpotent of order 2. Find $\phi, \psi : \mathbf{R}^2 \to \mathbf{R}^2$, each nilpotent of order 2, whose composition is idempotent.

D. Direct Sums, Quotients

In this section we discuss a generalization of the way we constructed the vector space $\mathbf{R}^2$ from $\mathbf{R}$. If V, W are vector spaces over a field k, we will construct a new vector space $V \oplus W$ called the *(external) direct sum* of V and W. The set of vectors in $V \oplus W$ is just the set of all ordered pairs (v, w) with $v \in V$ and $w \in W$. We add two such pairs coordinatewise

$$(v, w) + (x, y) = (v + x,\ w + y)$$

and for a scalar $r \in k$ we define

$$r(v,\ w) = (rv,\ rw).$$

The fact that these operations make $V \oplus W$ into a vector space is trivial. It satisfies the axioms for a vector space since V and W do.

The zero vector in $V \oplus W$ is $(0, 0)$, the first vector being the zero in V, the second that in W.

Just as we can include $\mathbf{R}$ into $\mathbf{R}^2$ as either the x-axis or y-axis, and we can project $\mathbf{R}^2$ onto either axis, we have four naturally defined linear maps.

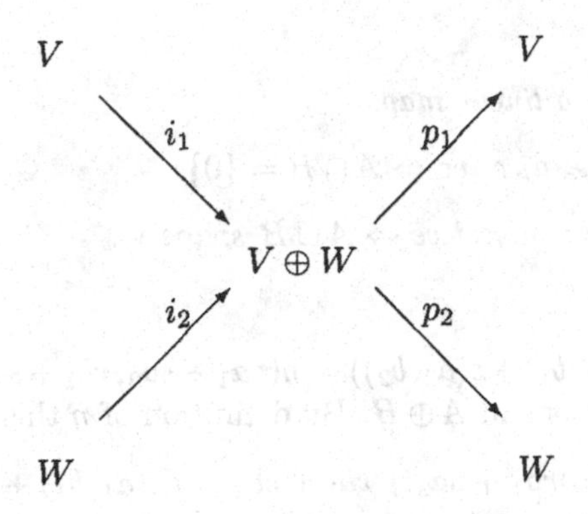

$$i_1(v) = (v, 0) \qquad p_1(x, y) = x$$
$$i_2(w) = (0, w) \qquad p_2(x, y) = y$$

We call i_1, i_2 *inclusions* and p_1, p_2 *projections.* Clearly $p_1 \circ i_1$ is the identity map on V and $p_2 \circ i_2$ is the identity map on W. It follows that i_1, i_2 are injective (linear maps) and p_1, p_2 are surjective (linear maps). Note also that $p_2 \circ i_1$ and $p_1 \circ i_2$ are zero maps. Actually, the image of i_1 is the kernel of p_2 and the image of i_2 is the kernel of p_1.

Proposition 12 *If* $\{v_1, \ldots, v_n\}$ *is a basis for* V *and* $\{w_1, \ldots, w_m\}$ *is a basis for* W *then*

$$\{(v_1, 0), \ldots, (v_n, 0), (0, w_1), \ldots, (0, w_m)\}$$

is a basis for $V \oplus W$.

Proof: Given $(v, w) \in V \oplus W$ we have

$$v = a_1 v_1 + \ldots + a_n v_n \text{ and}$$
$$w = b_1 w_1 + \ldots + b_m w_m \text{ uniquely.}$$

Then $(v, w) = a_1(v_1, 0) + \ldots + a_n(v_n, 0) + b_1(0, w_1) + \ldots + b_m(0, w_m)$, also uniquely. ∎

Corollary $\dim(V \oplus W) = \dim V + \dim W$

Next we consider when a vector space is, "somehow," the direct sum of two of its subspaces. So suppose A and B are subspaces of the vector space V (over k). Then A and B are vector spaces over k and we can construct $A \oplus B$ as above. We use the operation of vector addition in V to define a map

$$\eta : A \oplus B \to V,$$

$\eta(a, b) = a + b.$

Proposition 13

(i) *η is a linear map.*

(ii) *η is injective $\Leftrightarrow A \cap B = \{0\}$.*

(iii) *η is surjective $\Leftrightarrow A \cup B$ spans V.*

Proof:

(i) $\eta(r(a_1, b_1) + s(a_2, b_2)) = \eta(ra_1 + sa_2, rb_1 + sb_2)$ by definition of operations in $A \oplus B$. By definition of η this is

$$ra_1 + sa_2 + rb_1 + sb_2 = r\eta(a_1, b_1) + s\eta(a_2, b_2),$$

proving η is linear.

(ii) $\Leftarrow$ We suppose $A \cap B = \{0\}$ and $\eta(a, b) = 0$. That is $a + b = 0$. This implies $a = -b \in B$ and $b = -a \in A$ so $a = 0, b = 0$.

$\Rightarrow$ (Contrapositive proof) Suppose $c \in A \cap B$. If $c \neq 0$, we have $\eta(c, 0) = c$ and $\eta(0, c) = c$ but $(c, 0) \neq (0, c)$. Thus η is not monic.

(iii) $\Leftarrow$ Given $v \in V$, it is a finite linear combination of $a_1, \ldots, a_k$ in A and $b_1, \ldots, b_m$ in B. Add up (with coefficients) the part in A to get $a \in A$ and similarly get $b \in B$. Then

$$\eta(a, b) = a + b = v.$$

$\Rightarrow$ This is obvious. ∎

Definition When η (above) is an isomorphism, we say that V is an *internal direct sum* of A and B.

By Proposition 13, we conclude that V is the internal direct sum of A and B exactly when we can express every $v \in V$ uniquely as $v = a + b$ where $a \in A$, $b \in B$.

Examples

(1) $\mathbf{R}^2$ is the internal direct sum of

$$\begin{aligned} A &= \{(x, 0) \in \mathbf{R}^2\} \quad \text{and} \\ B &= \{(0, y) \in \mathbf{R}^2\}. \end{aligned}$$

(2) $\mathbf{R}^2$ is the internal direct sum of A (as in (1)) and $C = \{(x, x) \in \mathbf{R}^2\}$ because $A \cap C = \{0\}$ and $A \cup C$ spans $\mathbf{R}^2$. (Given $(p, q) \in \mathbf{R}^2$, we have

$$(p, q) = \underset{\in A}{(p - q, 0)} + \underset{\in C}{(q, q)}\,.)$$

There is some useful terminology for situations where we have three or more vector spaces and linear maps like, e.g.,

$$U \xrightarrow{\rho} V \xrightarrow{\sigma} W. \tag{6}$$

Definition We say the sequence (6) is *exact* at V if

$$\text{im } \rho = \ker \sigma \, .$$

Let us denote the trivial (= one-element) vector space by 0. Then if we assert that

$$0 \to U \xrightarrow{\rho} V$$

is exact at U, we are simply asserting that

$$\rho \text{ is monic.}$$

Similary, if we assert that

$$V \xrightarrow{\sigma} W \to 0$$

is exact at W, we are simply asserting that

$$\sigma \text{ is epic.}$$

Definition If

$$0 \to U \xrightarrow{\rho} V \xrightarrow{\sigma} W \to 0 \tag{7}$$

is an exact sequence of vector spaces (i.e., exact at U, at V, and at W), we say that (7) *splits* if there is a linear map γ (called a *splitting*)

$$V \xleftarrow{\gamma} W$$

such that $\sigma \circ \gamma$ is the identity on W.

Theorem 3 *For finite-dimensional vector spaces, the exact sequence (7) does split.*

Proof: Choose a basis $\{z_1, \ldots, z_m\}$ for W. By Proposition 9, given any m vectors $y_1, \ldots, y_m$ in V there is exactly one linear map $W \to V$ sending z_i to y_i, $i = 1, \ldots, m$. So if we choose

$$y_i \in \sigma^{-1}(z_i) \qquad i = 1, \ldots, m$$

(which we can do since σ is epic) and we set

$$\gamma(z_i) = y_i \, ,$$

then the resulting linear map clearly has $\sigma \circ \gamma = \text{id}$. ∎

Corollary *Given the exact sequence*

$$0 \to U \xrightarrow{\rho} V \xrightarrow{\sigma} W \to 0$$

of finite-dimensional vector spaces and any splitting $\gamma : W \to V$, then V is represented as an internal direct sum

$$V = \rho(U) \oplus \gamma(W).$$

Proof: Let $\{x_1, \ldots, x_n\}$ be a basis for U and $\{z_1, \ldots, z_m\}$ be a basis for W. Since $\rho(U) \cap \gamma(W) = \{0\}$ (by exactness at V), it will suffice to show that

$$\{\rho(x_1), \ldots, \rho(x_n), \gamma(z_1), \ldots, \gamma(z_m)\}$$

is a basis for V. Since ρ is injective, $\{\rho(x_1), \ldots, \rho(x_n)\}$ is linearly independent in V. Also $\{\gamma(z_1), \ldots, \gamma(z_m)\}$ is linearly independent in V (since their images under σ are linearly independent in $\gamma(W)$). Since $\rho(U) \cap \gamma(W) = \{0\}$, $\{\rho(x_1), \ldots, \rho(x_n), \gamma(z_1), \ldots, \gamma(z_m)\}$ is linearly independent in V.

By the rank theorem, $\dim V = n + m$ so that set is indeed a basis. ∎

Quotients

Let V be a vector space over a field k and let W be a subspace of V. We use W to define an equivalence relation on V. Namely, for $v, w \in V$, we define $v \sim w$ if $v - w \in W$. Now

(i) $v \sim v$, since $v - v = 0 \in W$.

(ii) $v \sim w \Rightarrow w \sim v$, since $v - w \in W \Rightarrow w - v = -(v - w) \in W$.

(iii) $v \sim w$ and $w \sim u \Rightarrow v \sim u$, since $v - u = (v - w) + (w - u)$ and W is closed under vector addition.

Thus $\sim$ is an equivalence relation.

The equivalence relation $\sim$ divides V into disjoint equivalence classes which are called *cosets.*

Proposition 14 *Let C be any one of these equivalence classes and let v be any element of C. Then*

$$C = W + v = \{w + v | w \in W\}.$$

Proof: By definition

$$C = \{u | u \sim v\} = \{u | u - v \in W\}.$$

For such a u, $u - v = w \in W$ so that $u = w + v$. So the proposition is clear. ∎

Thus each equivalence class is just "W translated by adding a fixed vector to each element of W." We have a schematic picture

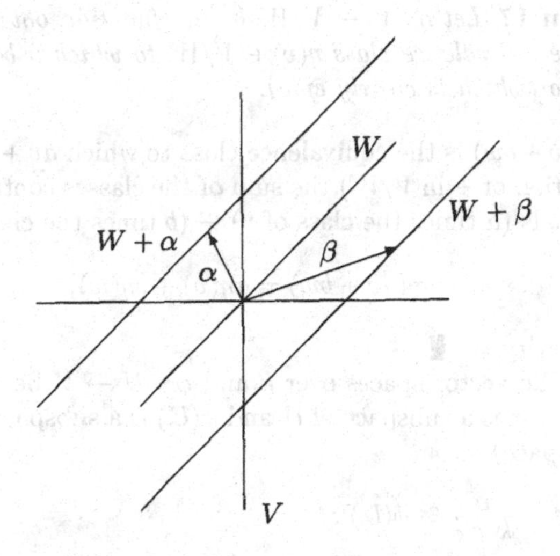

We denote the collection of equivalence classes by V/W and will now make V/W into a vector space, called the *quotient space* of V modulo W.

Definition Let C_1, C_2 be two equivalence classes. Choose any $v \in C_1$ and any $w \in C_2$ and define $C_1 + C_2$ to be the equivalence class containing the vector $v + w$. For any equivalence class C and any $r \in k$ we choose any $u \in C$ and define rC to be the equivalence class containing the vector ru.

Proposition 15 *These operations are well-defined, i.e., they are independent of our choices.*

Proof: Let $v' \in C_1$ and $w' \in C_2$. We need to show that the class containing $v + w$ is also the class containing $v' + w'$. But $v - v' = w_1 \in W$ and $w - w' = w_2 \in W$, so

$$(v + w) - (v' + w') = w_1 + w_2 \in W.$$

Thus addition is well-defined. If u' (as well as u) is in C, then $u - u' = w$ and $ru - ru' = r(u - u') = rw \in W$, so scalar multiplication is well-defined. ∎

Proposition 16 *With these operations V/W becomes a vector space over k.*

Proof: Exercise. ∎

Exercise. If $\dim V = n$ and $\dim W = m$, show that $\dim V/W = n - m$. (Hint: Let $\{w_1, \ldots, w_m\}$ be a basis for W and extend this to a basis $\{w_1, \ldots, w_m, v_{m+1}, \cdots, v_n\}$ for V. If $\bar{v}_i$ denotes the coset $W + v_i$, show that $\{\bar{v}_{m+1}, \ldots, \bar{v}_n\}$ is a basis for V/W.)

Proposition 17 *Let $\eta : V \to V/W$ be the function obtained by assigning to $v \in V$ the equivalence class $\eta(v) \in V/W$ to which v belongs. Then η is a linear map (which is clearly epic).*

Proof: $\eta(av + bw)$ is the equivalence class to which $av + bw$ belongs. This is (by definition of $+$ in V/W) the sum of the classes containing av and bw, and this equals (a times the class of v) $+$ (b times the class of w). That is,

$$\eta(av + bw) = a\eta(v) + b\eta(w),$$

proving linearity. ∎

Let U, V be vector spaces over k and $\phi : U \to V$ be a linear map. We know that $\ker \phi$ is a subspace of U and $\phi(U)$ is a subspace of V (and hence is a vector space).

Theorem 4 $\frac{U}{\ker \phi} \cong \phi(U)$.

Proof: We have

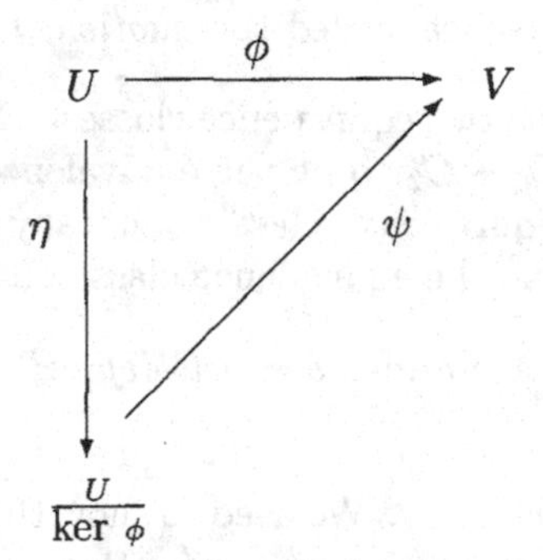

where ψ is defined as follows: for $u \in U$, $\eta(u) = u + \ker \phi$ and we set

$$\psi(\eta(u)) = \phi(u).$$

We claim: ψ is well-defined, ψ is linear, ψ is monic, and

$$\psi\left(\frac{U}{\ker \phi}\right) = \phi(U).$$

If $u + \ker\phi = u' + \ker\phi$, then $\eta(u) = \eta(u')$ so $\psi(\eta(u)) = \psi(\eta(u'))$ and ψ is well-defined.

$$\begin{aligned}\psi(a\eta(u) + b\eta(u')) &= \psi(\eta(au + bu')) \text{ (since } \eta \text{ is linear)}\\ &= \phi(au + bu') \text{ (by definition of } \psi)\\ &= a\phi(u) + b\phi(u') \text{ (}\phi \text{ is linear)}\\ &= a\psi(\eta(u)) + b\psi(\eta(u')) \text{ (definition of } \psi).\end{aligned}$$

Thus ψ is linear.

Clearly ψ is monic, for if $0 = \psi(\eta(u)) = \phi(u)$, then $u \in \ker\phi$ and $\eta(u) = 0$ in $\frac{U}{\ker\phi}$.

Finally, if $v = \phi(u)$ then $\psi(\eta(u)) = v$, so that $\psi(\frac{U}{\ker\phi}) = \phi(U)$. ∎

Exercises

(1) Show that V is the internal direct sum of the subspaces A and $B \Leftrightarrow$ every $v \in V$ can be written uniquely as $v = a + b$ where $a \in A$, $b \in B$ (i.e., A and B are complementary).

(2) Generalize the notions of external and internal direct sums to three or more summands. (Hint: The characterization of internal direct sum given in exercise (1) is easier to extend. Exercise (4), below, shows how to extend the definition given after Propositon 13.)

(3) If U and W are subspaces of the vector space V, show that

$$(U + W)/W \cong U/(U \cap W).$$

(4) Let $V_1, \ldots, V_n$ be subspaces of the vector space V and suppose that $V = V_1 + \cdots + V_n$. Show that if $V_i \cap (V_1 + \cdots + V_{i-1} + V_{i+1} + \cdots + V_n) = \{0\}$ for $i \in \{1, \ldots, n\}$, then V is the internal direct sum of $V_1, \ldots, V_n$.

(5) Let W be a subspace of the finite-dimensional vector space V. Show that there is a subspace U of V such that $V \cong U \oplus W$. (Hint: Consider the exact sequence

$$0 \to W \to V \to V/W \to 0$$

of finite-dimensional vector spaces.)

(6) Let W be the subspace of $\mathbf{R}^3$ spanned by $\{(1,1,0),(1,0,-1)\}$. Find a subspace $U \subseteq \mathbf{R}^3$ such that $\mathbf{R}^3 \cong U \oplus W$.

(7) Let V be finite-dimensional and suppose $V_1, \ldots, V_m$ are subspaces of V. Furthermore, suppose that B_i is a basis of V_i for each i. Show that $V = V_1 \oplus \cdots \oplus V_m \Leftrightarrow B = B_1 \cup \cdots \cup B_m$ is a basis of V.

The Linear Geometry of $\mathbf{R}^3$.

In the real vector space $\mathbf{R}^3$, a *line* is a coset of a 1-dimensional subspace of $\mathbf{R}^3$ (i.e., any line L is $L = v + V$ where V is a one-dimensional subspace of $\mathbf{R}^3$ and $v \in \mathbf{R}^3$). Similarly, a *plane* in $\mathbf{R}^3$ is a coset of a two-dimensional subset of $\mathbf{R}^3$. A *point* is a coset of the (trivial) zero-dimensional subspace $\{0\}$ of $\mathbf{R}^3$. The *dimension* of a coset S is the dimension of the corresponding subspace.

Note that the collection $\mathcal{S}$ of all cosets of $\mathbf{R}^3$ is partially ordered by inclusion.

Proposition (Prove) *Let* $\{S_\alpha\}_{\alpha \in A}$ *be a collection of cosets in* $\mathbf{R}^3$. *Prove that* $\bigcap_{\alpha \in A} S_\alpha$ *is either a coset of* $\mathbf{R}^3$ *or is empty.*

Definition Let S_1, S_2 be two cosets in $\mathbf{R}^3$. Their *join* is the intersection of all cosets $S \in \mathcal{S}$ such that $S_1 \subseteq S$ and $S_2 \subseteq S$, written $S_1 J S_2$.

Prove the following.

(1) Let V be a subspace of $\mathbf{R}^3$. Then

$$v + V = w + V \Leftrightarrow w \in v + V \Leftrightarrow -v + w \in V .$$

(2) If $w + W = v + V$, then $W = V$.

(3) $(w + W)J(v + V) = w + [-w + v] + W + V$, where $[-a + b]$ denotes the one-dimensional subspace generated by the vector $(-a + b)$.

(4) $(w + W) \cap (v + V) \neq \emptyset \Leftrightarrow (-w + v) \in W + V$.

Use (3) and (4) to prove.

Proposition *Let* $S = w + W$, $T = v + V$. *Then* $S \cap T = \emptyset \Leftrightarrow \dim(SJT) = \dim(W + V) + 1$.

Definition $w + W$ and $v + V$ are *parallel* if $W \subseteq V$ or $V \subseteq W$. (Note that a point is parallel to every coset).

Prove the following theorem.

Theorem *Let* S, T *be two cosets in* $\mathbf{R}^3$. *Then*

(1) $S \subseteq T \Rightarrow \dim S \leq \dim T$, *and equality implies* $S = T$.

(2) *If $S \cap T$ is not empty, then*

$$\dim(SJT) + \dim(S \cap T) = \dim S + \dim T .$$

(3) *If $S \cap T \neq \emptyset$, then S, T are parallel $\Leftrightarrow$ one contains the other.*

(4) *If $S \cap T = \emptyset$, then S, T are parallel $\Leftrightarrow$*

$$\dim(SJT) = \max\{\dim S, \dim T\} + 1.$$

Now prove the following theorem.

Theorem

(1) *The join of two distinct points is a line.*

(2) *The intersection of two nonparallel planes is a line.*

(3) *The join of two lines with a point of intersection is a plane.*

(4) *The intersection of two coplanar nonparallel lines is a point.*

(5) *The join of two distinct parallel lines is a plane.*

(6) *The join of a point with a line not containing this point is a plane.*

(7) *The intersection of a plane with a line not parallel to it is a point.*

E. Eigenvectors and Eigenvalues

(Part i)

In this section we are interested in linear maps of a vector space V into itself (i.e., elements of $\mathrm{End}(V)$).

Definition Let $\phi: V \to V$ be linear. A subspace W of V is said to be *ϕ-stable* if

$$\phi(W) \subseteq W.$$

Note that if W is ϕ-stable, then the restriction of ϕ to W is a linear map $\phi|_W : W \to W$.

Examples

(1) Let D be the real vector space of C^∞ functions from $[0,1]$ into $\mathbf{R}$. The set $E \subseteq D$ of *elementary functions* are those representable in closed form (i.e., start with $x, \log x, e^x, \cos x$, and $\arccos x$, and then construct new functions from these by using the standard algebraic operations and composition).

It is not hard to verify that E is a subspace of D. We looked earlier at the two linear maps

$$D \underset{\alpha}{\overset{\theta}{\leftrightarrows}} D$$

defined by $\alpha(f) = f'$ (i.e., differentiation) and

$$\theta(f)(x) = \int_0^x f(t)dt.$$

The subspace E is α-stable, but not θ-stable. By basic formulas for differentiation, if f is elementary, then so is its derivative f'. But if, for example, $f(t) = \sqrt{2\sin^2 t + \cos^2 t}$, then f is elementary, but $\theta(f)(x) = \int_0^x f(t)dt$ is not an elementary function. (It is an *elliptic* function.)

(2) A linear map $\rho : V \to V$ is said to be *idempotent* if $\rho \circ \rho = \rho$. For such a linear map, $\rho(V)$ is ρ-stable. Indeed suppose $y \in \rho(V)$, i.e., $y = \rho(v)$ for some $v \in V$. Then

$$\rho(y) = \rho(\rho(v)) = \rho(v) = y$$

so that ρ restricted to $\rho(V)$ is the identity map.

(3) Let $\phi : \mathbf{R^2} \to \mathbf{R^2}$ be defined by

$$\phi(x_1, x_2) = (x_1 + x_2, x_1 + x_2).$$

Then the *diagonal* $D = \{(x, x) \in \mathbf{R^2}\}$ is ϕ-stable because $\phi(x, x) = (2x, 2x) \in D$.

(4) Let D be the real vector space of C^∞ functions and let $\phi : D \to D$ be differentiation. Let $W = \{re^x | r \in \mathbf{R}\}$. This is a one-dimensional linear subspace of D which is ϕ-stable. Indeed, ϕ is the identity map on W.

The idea of finding ϕ-stable subspaces W of V is to study ϕ by observing its action on the ϕ-stable subspaces. The point is that it is easier to understand linear mappings on lower-dimensional spaces. In particular, we noted earlier that a linear map $\alpha : \ k \to k$ (k a field) is completely determined by knowing its value on one element of k, e.g., $\alpha(1)$. So we make the following definition.

Definition A nonzero vector $v \in V$ is an *eigenvector* for $\phi: V \to V$ if Span(v) is ϕ-stable.

A more customary way of phrasing this definition is to say that $v \neq 0$ is an eigenvector for ϕ if there exists some $\lambda \in k$ such that

$$\phi(v) = \lambda v.$$

Notice that in this statement the choice $\lambda = 0$ is not excluded. Thus every nonzero vector $v \in \ker \phi$ is an eigenvector for ϕ. If v is an eigenvector for ϕ then the (unique) λ such that

$$\phi(v) = \lambda v$$

is called an *eigenvalue* for ϕ corresponding to v.

Proposition 18 *If v is an eigenvector for ϕ, and r is any nonzero element of k, then rv is also an eigenvector for ϕ with the same eigenvalue as v.*

Proof: Let λ be the eigenvalue of v. Then

$$\phi(rv) = r\phi(v) = r(\lambda v) = \lambda(rv). \quad \blacksquare$$

This suggests that, for a linear map $\phi: V \to V$, the notion of eigenvalue may be more fundamental than the notion of eigenvector. So for $\lambda \in k$ we let

$$V(\lambda) = \{v \in V | \phi(v) = \lambda v\}.$$

Then each $V(\lambda)$ contains the zero vector, and each nonzero vector in $V(\lambda)$ is an eigenvector of ϕ having λ as eigenvalue. Note that $V(0) = \ker \phi$. Also note that each $V(\lambda)$ is a subspace of V called the *eigenspace belonging to* λ. (Of course, we expect that most $V(\lambda)$ will consist of the zero vector only.)

The dimension of $V(\lambda)$ is the *geometric multiplicity* of λ.

Proposition 19 *If $\lambda, \mu \in k$ and $\lambda \neq \mu$, then*

$$V(\lambda) \cap V(\mu) = \{0\}.$$

Proof: If $v \in V(\lambda) \cap V(\mu)$, then

$$\phi(v) = \lambda v = \mu v$$

so that

$$(\lambda - \mu)v = 0.$$

Thus if $\lambda \neq \mu$, then $v = 0$. $\blacksquare$

An Important Example:

There exists a linear map $\phi : \mathbf{R^2} \to \mathbf{R^2}$ such that $V(\lambda) = \{0\}$ for each $\lambda \in \mathbf{R}$.

Let $e_1 = (1,0)$ and $e_2 = (0,1)$. Define ϕ by $\phi(e_1) = e_2, \phi(e_2) = -e_1$. If $v = ae_1 + be_2$ is in $V(\lambda)$, then

$$\begin{aligned} &\phi(v) = a\phi(e_1) + b\phi(e_2) = \lambda v = \lambda(ae_1 + be_2) \\ \text{so} \quad &ae_2 + b(-e_1) = \lambda ae_1 + \lambda be_2. \end{aligned}$$

Thus $\lambda b = a$ and $\lambda a = -b$. So $\lambda(\lambda b) = \lambda a = -b$. Since $\lambda^2 \neq -1$ in $\mathbf{R}$ we must have $b = 0$. But then $v = ae_1$ and $\phi(v) = a\phi(e_1) = ae_2 = \lambda ae_1$ implying that $a = 0$ as well as $b = 0$. So $v = 0$. The point is that ϕ rotates the plane $\mathbf{R^2}$ by 90° and the only vector which goes to a scalar multiple of itself is the zero vector.

Let $\phi \in \text{End}(V)$ and $\psi \in GL(V)$ $(\subseteq \text{End}(V))$.

Theorem 5 *If v is an eigenvector for ϕ with eigenvalue λ, then $\psi(v)$ is an eigenvector for $\psi\phi\psi^{-1}$ with eigenvalue λ.*

Proof: $\psi\phi\psi^{-1}(\psi(v)) = \psi\phi(v) = \psi(\lambda v) = \lambda\psi(v)$. ∎

Observation:

Suppose V has a basis $\{v_1, \ldots, v_n\}$ consisting of eigenvectors of ϕ. If $\lambda_1, \ldots, \lambda_n$ are the corresponding eigenvalues, then ϕ is simply described. If $v = a_1v_1 + \ldots + a_nv_n$ then

$$\phi(v) = a_1\lambda_1v_1 + \ldots + a_n\lambda_nv_n.$$

Exercises

(1) Show that $V(\lambda)$ is, indeed, a subspace of V for each $\lambda \in k$.

(2) Let $\{e_1, e_2\}$ be the standard basis for $\mathbf{R^2}$. Find the eigenvalues and corresponding eigenvectors for $\phi : \mathbf{R^2} \to \mathbf{R^2}$ defined by $\phi(e_1) = 2e_1 + e_2$, $\phi(e_2) = 2e_1 + 3e_2$. What is the geometric multiplicity of each eigenvalue?

(3) Let $\{e_1, e_2, e_3\}$ be the standard basis for $\mathbf{R^3}$. Find the eigenvalues and corresponding eigenvectors for $\phi : \mathbf{R^3} \to \mathbf{R^3}$ defined by $\phi(e_1) = e_1$, $\phi(e_2) = e_1 + e_2$, $\phi(e_3) = e_3$. What is the geometric multiplicity of each eigenvalue?

(4) Recall that we defined $\phi : \mathbf{R^2} \to \mathbf{R^2}$ in the important example by $\phi(e_1) = e_2$, $\phi(e_2) = -e_1$. Now suppose that we consider $\{e_1, e_2\}$ as a basis for $\mathbf{C}^2$ and define $\phi : \mathbf{C}^2 \to \mathbf{C}^2$ by $\phi(e_1) = e_2$, $\phi(e_2) = -e_1$. Does this change matters? That is, does ϕ have any eigenvalues now, and if so, what are the corresponding $V(\lambda)$'s?

(5) Given $\phi \in \text{End}(V)$, show that 0 is an eigenvalue for $\phi \Leftrightarrow \ker \phi \neq \{0\}$.

(6) Suppose that λ is an eigenvalue for the isomorphism $\phi \in \text{End}(V)$. Show that λ^{-1} is an eigenvalue for ϕ^{-1}.

(7) Given $\phi, \psi \in \text{End}(V)$, show that $\phi\psi$ and $\psi\phi$ have the same eigenvalues. (Hint: consider the cases $\lambda = 0$ and $\lambda \neq 0$ separately.)

(8) Suppose that the vector space V is the direct sum of the subspaces $V_1, \ldots, V_m$. Also, suppose that we have endomorphisms $\phi_1, \ldots, \phi_m$ where $\phi_i \in \text{End}(V_i)$. Given $v \in V$, we can write it (uniquely) as $v = v_1 + \cdots + v_m$ where $v_i \in V_i$. If we define

$$\phi = \phi_1 \oplus \cdots \oplus \phi_m$$

by $\phi(v) = \phi_1(v_1) + \cdots + \phi_m(v_m)$, show that ϕ is an endomorphism of V. Furthermore, show that each V_i is ϕ-stable and that $\phi|_{V_i} = \phi_i$.

(9) Suppose we have $\phi \in \text{End}(V)$ and ϕ-stable subspaces $V_1, \ldots, V_m$ such that $V = V_1 \oplus \cdots \oplus V_m$. Show that $\phi(V) = \phi(V_1) \oplus \cdots \oplus \phi(V_m)$.

F. Dual Spaces

Let V be a vector space over a field k, and let V^* be the set of all linear maps

$$\phi : V \to k.$$

Then V^* becomes a vector space by defining

$$(r\phi)(v) = r\phi(v) \quad \text{and}$$
$$(\phi + \psi)(v) = \phi(v) + \psi(v)$$

for $r \in k$ and $\phi, \psi \in V^*$.

Definition This vector space V^* is called the *dual space* of V, and its elements are called *functionals*.

Examples

(1) Let $V = k^n$ and let $\phi : k^n \to k$ be projection on the ith factor; i.e.,

$$\phi(x_1, \ldots, x_n) = x_i \, .$$

Then ϕ is a functional.

(2) Let V be finite-dimensional and choose a basis $\{v_1, \ldots, v_n\}$ so that $v \in V$ can be written as $v = x_1v_1 + \cdots + x_nv_n$ for $x_i \in k$. Then for any $(a_1, \ldots, a_n) \in k^n$ we get a functional defined by

$$\phi(v) = a_1x_1 + \cdots + a_nx_n .$$

In Proposition 20, we shall see that the converse is also true. That is, any $\phi \in V^*$ can be written in this form once we choose a basis for V.

(3) Let $V = C([0,1])$ be the vector space of real-valued continuous functions on $[0,1]$. Then

$$\phi(f) = \int_0^1 f(t)dt$$

is a functional.

(4) Let V be any vector space (possibly infinite-dimensional) and let B be a basis for V. If we choose $v \in B$, then there is a unique functional v^* such that $v^*(v) = 1$ and $v^*(w) = 0$ for any $w \in B - \{v\}$.

<u>Exercise</u>. Verify these examples.

Proposition 20 *If V is finite-dimensional, then* $\dim(V^*) = \dim(V)$ *so that* $V \cong V^*$.

Proof: Choose a basis $\alpha = \{v_1, \ldots, v_n\}$ for V and define $v_1^*, \ldots, v_n^* \in V^*$ by

$$v_i^*(v_j) = \delta_{ij} .$$

We claim that $\alpha^* = \{v_1^*, \ldots, v_n^*\}$ is a basis of V^* (called the *dual basis* of α).

First, α^* spans V^*. For if $\phi \in V^*$, then we have

$$\phi = \phi(v_1)v_1^* + \cdots + \phi(v_n)v_n^*$$

(just apply both sides of this expression to any v_i to get $\phi(v_i)$).

Next, α^* is linearly independent. For if we have scalars $c_1, \ldots, c_n$ such that

$$c_1v_1^* + \cdots + c_nv_n^* = 0 \in V^* ,$$

then $0 = (c_1v_1^* + \cdots + c_nv_n^*)(v_i) = c_i$ for each i.

We get an isomorphism $\psi_\alpha : V \to V^*$ by defining

$$\psi_\alpha(v_i) = v_i^* . \quad ∎$$

We remark that the isomorphism $V \cong V^*$ depends strongly on our choice of basis (i.e., a different basis β for V would lead to a different dual basis β^* and a different isomorphism ψ_β). In fact, there is no obvious way to identify V to V^* unless we choose a basis.

Since we use the term "dual space", it seems reasonable to ask if, somehow, V is also the dual of V^*. We will see that when V is finite-dimensional, we have a natural way to identify V with the dual space $(V^*)^*$ of V^*. Here, "natural" means "basis free." That is, our isomorphism of V onto $(V^*)^*$ will not require us to choose a basis. We still have a natural map of V into $(V^*)^*$ when V is infinite-dimensional, but it is only monic in this case.

We now define our map

$$\rho : V \to (V^*)^*$$

as follows. Given $v \in V$, we want $\rho(v)$ to be a functional on V^*. So given $\phi \in V^*$, we define

$$\rho(v)(\phi) = \phi(v) .$$

Exercise. Show that $\rho(v)$ is, indeed, a functional on V^*.

Proposition 21 *$\rho : V \to (V^*)^*$ is a linear map.*

Proof: Choose $v_1, v_2 \in V$ and $r \in k$ and $\phi \in V^*$. Then we have

$$\rho(rv_1)(\phi) = \phi(rv_1) = r\phi(v_1) = r\rho(v_1)(\phi)$$

and

$$\begin{aligned} \rho(v_1 + v_2)(\phi) = \phi(v_1 + v_2) = \phi(v_1) + \phi(v_2) &= \rho(v_1)(\phi) + \rho(v_2)(\phi) \\ &= (\rho(v_1) + \rho(v_2))(\phi). \end{aligned}$$

Since ϕ is arbitrary, we must have

$$\rho(rv_1) = r\rho(v_1) \quad \text{and} \quad \rho(v_1 + v_2) = \rho(v_1) + \rho(v_2) . \quad \blacksquare$$

We are going to show that the linear map

$$\rho : V \to (V^*)^*$$

is monic (even when V is infinite-dimensional). This will use a logical principle known as *Zorn's lemma*, so we make a slight detour to discuss that.

A nonempty set S is *partially ordered* if for some pairs $(a, b) \in S \times S$, there is a relation, written $a \leq b$, which satisfies the properties

(1) $a \leq a$,

(2) if $a \leq b$ and $b \leq c$, then $a \leq c$, and

(3) if $a \leq b$ and $b \leq a$, then $a = b$.

The elements $a, b \in S$ are said to be *comparable* if either $a \leq b$ or $b \leq a$. Note that we do not require all pairs of elements to be comparable in a partially ordered set; however, if $a, b \in S$ are always comparable, then S is *simply ordered* (or *linearly ordered* or *totally ordered*).

A partially ordered set S has a *maximal element* a_0 if $a \leq a_0$ for all $a \in S$ which are comparable to a_0. An *upper bound* for the nonempty subset $T \subseteq S$ is an element $b_0 \in S$ such that $b \leq b_0$ for all $b \in T$.

Zorn's Lemma: Let S be a partially ordered set, and suppose that each simply ordered subset of S has an upper bound in S. Then S contains a maximal element.

Now let V be a vector space over the field k. We will use Zorn's lemma to show that any nonzero element $v \in V$ is part of a basis for V. (In particular, V always has a basis, as claimed earlier.) Let S be the collection of all subsets of V which are linearly independent and contain v. We partially order S by inclusion: for $a, b \in S$ we write $a \leq b$ to mean $a \subseteq b$.

Now let T be a simply ordered subset of S, and let U be the union of all the elements (sets) in T. Clearly $v \in U$, and we claim that $U \in S$, which will show that U is an upper bound for T. So we just need to show that U is a linearly independent set. Suppose $u_1, \ldots, u_m$ are elements of U with $a_1u_1 + \cdots + a_mu_m = 0$ where $a_1, \ldots, a_m \in k$. Then, since all $u_1, \ldots, u_m$ are elements of some $\tau \in T$, and every $\tau \in T$ is a linearly independent set in S, we must have all $a_i = 0$.

Thus S satisfies the hypotheses of Zorn's lemma, and we conclude that S has a maximal element $s_0 \in S$. Now s_o is a linearly independent set in V and it contains v. If s_0 does not span V, we can choose $w \in V$ outside the span of s_0. But then $s_0 \cup \{w\}$ is a linearly independent set containing v, contradicting the maximality of s_0. Thus we have proven

Theorem 6 *Let V be a vector space over k and suppose $v \in V$ is any nonzero element. Then V has a basis containing v as one element.*

Corollary Given the nonzero element $v \in V$, there exists a linear map $\phi : V \to k$ such that $\phi(v) = 1$.
Proof: Take a basis for V containing v and apply example (4). ∎

Proposition 22 *$\rho : V \to (V^*)^*$ is monic.*

Proof: Suppose ρ is not monic. Then for some nonzero $v \in V$, we have $\rho(v) = 0$. This means $\rho(v)$ is the zero map $V^* \to k$; that is, for any $\phi \in V^*$ we have $\rho(v)(\phi) = \phi(v) = 0$. But, by the corollary, some ϕ maps v to 1, a contradiction. ∎

Proposition 23 *If V is finite-dimensional, then ρ is an isomorphism.*

Proof: We know that $\dim V = \dim(V^*) = \dim((V^*)^*)$, so the monic map ρ must also be epic. ∎

Exercises

(1) Show that the set of integers $\mathbf{Z}$ equipped with the standard order is simply ordered but has no maximal element.

(2) Consider the subset of $\mathbf{R}^2$

$$S = \{(x_1, x_2) \mid x_2 \leq 0\}$$

and define a relation on S by $(x_1, x_2) \leq (y_1, y_2) \Leftrightarrow x_1 = y_1$ and $x_2 \leq y_2$. Show that S is partially ordered by this relation. Show also that S has infinitely many maximal elements.

(3) Given the basis $\{(1,0,0), (1,-1,0), (2,0,1)\}$ of $\mathbf{R}^3$, find its dual basis.

(4) Generalize Theorem 6 as follows: let V be a vector space over k and let A be a linearly independent subset of V. Show that A can be extended to a basis for V.

(5) Let W be a subspace of the vector space V. If $\phi \in W^*$, show that we can find a $\tilde{\phi} \in V^*$ such that $\tilde{\phi}\Big|_W = \phi$.

(6) Given the subset W of the vector space V, we say that $\phi \in V^*$ *annihilates* W if $\phi(w) = 0$ for every $w \in W$. We call

$$A(W) = \{\phi \in V^* \mid \phi \text{ annihilates } W\}$$

the *annihilator* of W.

Show that $A(W)$ is a subspace of V^*. Show also that if $W' \subseteq W$, then $A(W) \subseteq A(W')$.

(7) If W is a subspace of the finite-dimensional vector space V, show that

$$W^* \cong V^*/A(W)\,.$$

Conclude that $\dim(A(W)) = \dim V - \dim W$. (Hint: Define a map

$$\psi : V^* \to W^*$$

by $\psi(\phi) = \phi|_W$ and show that ψ is surjective with $\ker \psi = A(W)$. Now apply Theorem 4.)

(8) Use exercise (7) to show that $W = A(A(W))$. Note that we must identify V and $(V^*)^*$ for this to make sense. (Hint: First show that $W \subseteq A(A(W))$.)

(9) Let U and W be subspaces of the (not necessarily finite-dimensional) vector space V. Show that $A(U + W) = A(U) \cap A(W)$. If V is finite-dimensional, show also that $A(U \cap W) = A(U) + A(W)$.

(10) If $V = U \oplus W$, show that $V^* = A(W) \oplus A(U)$.

Chapter II

Matrices and Determinants

A. Matrices

We have already noticed that a linear map $\phi : k \to k$ is completely determined by its value on one element of k, e.g., on 1. Also we noted that a linear map

$$\phi : k^2 \to k$$

is determined by $\phi(1,0)$ and $\phi(0,1)$. We will now see how a linear map

$$\phi : V \to W$$

(V, W being vector spaces over k of dimensions n, m) is determined by nm elements of k.

Let $\{v_1, \ldots, v_n\}$ be a basis for V and $\{w_1, \ldots, w_m\}$ be a basis for W. Then (by Proposition 8 of Chapter I) each $\phi(v_1)$ is a unique linear combination of $w_1, \ldots, w_m$. We write this as

$$\phi(v_i) = a_{i1}w_1 + a_{i2}w_2 + \cdots + a_{im}w_m \,. \tag{1}$$

Also, we know that ϕ is uniquely determined by the n vectors (in W) $\phi(v_1), \ldots, \phi(v_n)$. Thus the rectangular array

$$M(\phi) = \begin{pmatrix} a_{11} & a_{12} & \ldots & a_{1m} \\ a_{21} & a_{22} & \ldots & a_{2m} \\ \vdots & & & \\ a_{n1} & a_{n2} & \ldots & a_{nm} \end{pmatrix} \tag{2}$$

of nm elements of k determines ϕ, as long as we know the bases $\{v_1, \ldots, v_n\}$, $\{w_1, \ldots, w_m\}$. This rectangular array of elements of k is called a *matrix* and the one shown has n *rows* and m *columns*; it is called an $n \times m$ *matrix*; and $M(\phi)$ is called the *matrix representation* of ϕ with respect to the bases $\{v_1, \ldots, v_n\}$ and $\{w_1, \ldots, w_m\}$. We sometimes write

$$M(\phi) = (a_{ij})$$

meaning that we will denote the element in the ith row and j^{th} column by a_{ij}, and we call it the i, j *entry* of (a_{ij}).

With the given bases for V and W and the matrix $M(\phi)$ we can describe ϕ quite explicitly. Given $x \in V$, we have

$$x = x_1 v_1 + \cdots + x_n v_n \qquad \text{uniquely.}$$

Since ϕ is linear, we have

$$\phi(x) = x_1 \phi(v_1) + \cdots + x_n \phi(v_n).$$

Then, by using (1), we have

$$\begin{aligned} \phi(x) &= x_1(a_{11}w_1 + \cdots + a_{1m}w_m) + \cdots + x_n(a_{n1}w_1 + \cdots + a_{nm}w_m) \\ &= (x_1 a_{11} + x_2 a_{21} + \cdots + x_n a_{n1})w_1 + \ldots \\ &+ (x_1 a_{1m} + \cdots + x_n a_{nm})w_m . \end{aligned}$$

In particular, we see that the coefficient of w_j in the unique expression for $\phi(x)$ is

$$(3) \qquad y_j = x_1 a_{1j} + x_2 a_{2j} + \cdots + x_n a_{nj}.$$

This is the element of k obtained by multiplying the *row* $(x_1, x_2, \ldots x_n)$ by the *column* $\begin{pmatrix} a_{1j} \\ a_{2j} \\ \vdots \\ a_{nj} \end{pmatrix}$ term-by-term and then adding up.

Examples

First we give some simple examples where $V = W = R^2$ and we use the basis $\{e_1 = (1, 0), e_2 = (0, 1)\}$ for both V and W.

	ϕ	$M(\phi)$	Remarks
(1)	$\phi(e_1) = 0, \phi(e_2) = 0$	$\begin{pmatrix} 0 & 0 \\ 0 & 0 \end{pmatrix}$	The zero linear map
(2)	$\phi(e_1) = e_2, \phi(e_2) = 0$	$\begin{pmatrix} 0 & 1 \\ 0 & 0 \end{pmatrix}$	A nilpotent map
(3)	$\phi(e_1) = 0, \phi(e_2) = e_1$	$\begin{pmatrix} 0 & 0 \\ 1 & 0 \end{pmatrix}$	Nilpotent
(4)	$\phi(e_1) = 0, \phi(e_2) = e_2$	$\begin{pmatrix} 0 & 0 \\ 0 & 1 \end{pmatrix}$	Idempotent
(5)	$\phi(e_1) = e_1 + e_2, \phi(e_2) = e_1 - e_2$	$\begin{pmatrix} 1 & 1 \\ 1 & -1 \end{pmatrix}$	
(6)	$\phi(e_1) = e_2, \phi(e_2) = -e_1$	$\begin{pmatrix} 0 & 1 \\ -1 & 0 \end{pmatrix}$	rotation by $\pi/2$
(7)	$\phi(e_1) = e_2, \phi(e_2) = e_1$	$\begin{pmatrix} 0 & 1 \\ 1 & 0 \end{pmatrix}$	reflection in line
(8)	$\phi(e_1) = -e_1, \phi(e_2) = -e_2$	$\begin{pmatrix} -1 & 0 \\ 0 & -1 \end{pmatrix}$	reflection in origin
(9)	$\phi(e_1) = e_1 + e_2, \phi(e_2) = e_1 + e_2$	$\begin{pmatrix} 1 & 1 \\ 1 & 1 \end{pmatrix}$	

<u>Exercise</u>. Which of these nine are isomorphisms?

An example in $V = W = R^3$: let

$$\{v_1 = (1,0,0), \quad v_2 = (1,1,1), \quad v_3 = (0,1,0)\}$$

be the basis for V and

$$\{w_1 = (1,0,0), \quad w_2 = (0,1,0), \quad w_3 = (0,0,1)\}$$

be the basis for W. Define ϕ by

$$\phi(v_1) = w_1 + w_2, \quad \phi(v_2) = w_2, \quad \phi(v_3) = w_1 - w_3.$$

<u>Exercise</u>. Write out $M(\phi)$ and try to decide if ϕ is an isomorphism.

We have seen that given bases for V and W, a linear map ϕ determines a matrix $M(\phi)$ (which will, in turn, determine ϕ). Suppose now we are given bases $\{v_1, \ldots, v_n\}$ for V and $\{w_1, \ldots, w_m\}$ for W, and we are also given an $n \times m$ matrix

$$Z = (z_{ij}) , \qquad z_{ij} \in k.$$

Then (z_{ij}) determines a linear map $\psi : V \to W$ by

$$\psi(v_i) = z_{i1}w_1 + \cdots + z_{im}w_m.$$

This leads to an interesting point. Suppose we have vector spaces V, W, U of dimensions n, m, p respectively and have linear maps

$$V \xrightarrow{\phi} W \xrightarrow{\psi} U.$$

Then the composition $\psi \circ \phi$ is a linear map from V to U. If we have chosen bases

$$\begin{aligned} \{v_1, \ldots, v_n\} &\text{ for } V \\ \{w_1, \ldots, w_m\} &\text{ for } W \\ \{u_1, \ldots, u_p\} &\text{ for } U, \end{aligned}$$

then we have matrices $M(\phi)$, $M(\psi)$, and $M(\psi \circ \phi)$ relative to these bases. Since ϕ and ψ determine $\psi \circ \phi$, it should be that $M(\phi)$ and $M(\psi)$ determine $M(\psi \circ \phi)$. We now explore how this works.

$$\text{Let} \quad \begin{aligned} M(\phi) &= (a_{ij}) \\ M(\psi) &= (b_{ij}) \\ M(\psi \circ \phi) &= (c_{ij}). \end{aligned}$$

Then we have

$$\begin{aligned} \phi(v_i) &= a_{i1} w_1 + \cdots + a_{im} w_m \\ \psi(w_\ell) &= b_{\ell 1} u_1 + \cdots + b_{\ell p} u_p \\ (\psi \circ \phi)(v_i) &= c_{i1} u_1 + \cdots + c_{ip} u_p \, . \end{aligned}$$

Using the first two expressions we can calculate the coefficient c_{ij} of u_j in the third expression. We have

$$\psi(\phi(v_i)) = a_{i1} \psi(w_1) + \cdots + a_{im} \psi(w_m)$$

and we see that

$$c_{ij} = a_{i1} b_{1j} + a_{i2} b_{2j} + \cdots + a_{im} b_{mj} \, . \tag{4}$$

That is, we get the ij entry in $M(\psi \circ \phi)$ by multiplying the ith row of $M(\phi)$ term-by-term with the jth column of $M(\psi)$ and summing the resulting m terms.

So we use this as our definition of how to *multiply* an $n \times m$ matrix with an $m \times p$ matrix to obtain an $n \times p$ matrix, i.e., $(a_{ij}) \cdot (b_{ij}) = (c_{ij})$ where c_{ij} is given by (4).

Example

$$\underset{2 \times 3}{\begin{pmatrix} 1 & 2 & 3 \\ 0 & -4 & 7 \end{pmatrix}} \underset{3 \times 2}{\begin{pmatrix} 2 & 1 \\ 3 & 0 \\ -3 & 3 \end{pmatrix}} = \underset{2 \times 2}{\begin{pmatrix} -1 & 10 \\ -33 & 21 \end{pmatrix}}$$

$$n = 2, \quad m = 3, \quad p = 2$$

We have also proven the following.

Theorem *With ϕ and ψ as above, we have*

$$M(\psi \circ \phi) = M(\phi) \cdot M(\psi).$$

Notice that formula (3) for calculating $\phi(x)$ is just a special case of matrix multiplication. We had bases $\{v_1, \ldots, v_n\}$ for V and $\{w_1, \ldots, w_m\}$ for W. Using this basis for V we can think of

$$x = x_1 v_1 + \cdots + x_n v_n$$

as a $1 \times n$ matrix $(x_1, x_2, \ldots, x_n)$ called the *coordinate vector* of x with respect to the basis $\{v_1, \ldots, v_n\}$. Then (3) tells us we get $\phi(x)$ (relative to $\{w_1, \ldots, w_m\}$) as the $1 \times m$ matrix $(y_1, y_2, \ldots, y_m)$:

$$(x_1, x_2, \ldots, x_n) \begin{pmatrix} a_{11} & \ldots & a_{1m} \\ \vdots & & \vdots \\ a_{n1} & \ldots & a_{nm} \end{pmatrix} = (y_1, y_2, \ldots, y_m).$$

Example

If $W = k$, then $\phi : V \to k$ is simply a functional on V and we have

$$\phi(x) = (x_1, \ldots, x_n) \begin{pmatrix} a_1 \\ \vdots \\ a_n \end{pmatrix} = a_1 x_1 + \cdots + a_n x_n$$

as in example (2) of section I F.

An Important Observation:

By now, it has probably become apparent that there is a slight discrepancy in our notation; i.e., we write functions on the left (e.g., $x \mapsto \phi(x)$), but when we represent a linear map by a matrix, we write it on the right (e.g., $(x_1, \ldots, x_n) \mapsto (x_1, \ldots, x_n)M(\phi)$). This difference is due to the fact that we will want to consider linear maps over the quaternions in Chapter V. We will see that the quaternions are noncommutative, so our choice to define scalar multiplication on the left (i.e., (scalar)(vector)) forces us to write a linear map as matrix multiplication on the right.

The only real difficulty caused by this choice arises when we compose maps: if we want to find $M(\psi \circ \phi)$, we must compute $M(\phi) \cdot M(\psi)$ instead of $M(\psi) \cdot M(\phi)$.

It is left as an exercise to show that this difficulty disappears if we write vectors as column matrices and multiply by the matrix of a linear map on the left.

Square Matrices

Let V be an n-dimensional vector space. We will be considering linear maps $V \to V$, i.e., elements of End(V); these correspond to *square* $(n \times n)$ *matrices.*

Proposition 1 *Let* $\{v_1, \ldots, v_n\}$ *and* $\{w_1, \ldots, w_n\}$ *be any two bases for* V. *Then the linear map* $\phi : V \to V$ *defined by* $\phi(v_i) = w_i$ *is an isomorphism.*

Proof: By the rank theorem, if ϕ is monic, it will also be epic and hence an isomorphism. So suppose $\phi(v) = 0$. Write $v = a_1 v_1 + \ldots + v_n v_n$ and then

$$0 = \phi(v) = a_1\phi(v_1) + \cdots + a_n\phi(v_n) = a_1 w_1 + \cdots + a_n w_n.$$

Since $\{w_1, \ldots, w_n\}$ is a basis, each $a_i = 0$ and thus $v = 0$. ∎

Notation:

If we use the same basis $\alpha = \{v_1, \ldots, v_n\}$ for V as domain and V as range, we denote the matrix for $\rho \in \text{End}(V)$ by $M_\alpha(\rho)$.

If $I : V \to V$ is the identity map, then for any basis α we have $M_\alpha(I) = (\delta_{ij})$ where

$$\delta_{ij} = 1 \quad \text{if } i = j, \quad \text{and}$$
$$\delta_{ij} = 0 \quad \text{if } i \neq j .$$

This is called the *identity matrix* I. Clearly $I \circ \rho = \rho \circ I = \rho$ for any $\rho \in \text{End}(V)$, and for any $n \times n$ matrix A, $AI = IA = A$.

Definition The $n \times n$ matrix A is a *nonsingular* if there exists an $n \times n$ matrix B such that $AB = BA = I$; B is called the *inverse* of A and is usually written as A^{-1}.

Proposition 2 *If* $\rho \in \text{End}(V)$ *is an isomorphism, then any* $M_\alpha(\rho)$ *has an inverse, namely,* $M_\alpha(\rho^{-1})$.

Proof: $I = M_\alpha(\rho^{-1} \circ \rho) = M_\alpha(\rho)M_\alpha(\rho^{-1})$. ∎

Proposition 3 *A matrix* A *is nonsingular* $\Leftrightarrow$ *any corresponding* ϕ *is an isomorphism.*

Proof: $\Leftarrow$ comes from Proposition 2.

$\Rightarrow$ If $A = M_\alpha(\phi)$ has an inverse $A^{-1} = B = (b_{ij})$, where $\alpha = \{v_1, \ldots, v_n\}$ is some basis for V, we define $\psi : V \to V$ by

$$\psi(v_i) = b_{i1}v_1 + \cdots + b_{in}v_n$$

and note that $\psi \circ \phi$ and $\phi \circ \psi$ are represented by I. ∎

Definition Two $n \times n$ matrices A, B are *similar* (or *conjugate*) if there exists a nonsingular $n \times n$ matrix C such that

$$B = CAC^{-1}.$$

It is an exercise to show that similarity is an equivalence relation. (It seems that "conjugate" is a much better word than "similar" for this relationship.)

An obvious question to ask at this point is the following: how does the matrix representing a given endomorphism of V change when we choose a different basis for V? Fortunately, this is easy to answer.

Proposition 4 *Let $\phi \in \text{End}(V)$. Let $A = M_\alpha(\phi)$ and $B = M_\beta(\phi)$ for two bases $\alpha = \{v_1, \ldots, v_n\}$, $\beta = \{w_1, \ldots, w_n\}$. Then A and B are similar matrices.*

Proof: If $A = (a_{ij})$ and $B = (b_{ij})$, then by (1) we have

$$\begin{aligned} \phi(w_i) &= \sum_{j=1}^{n} b_{ij} w_j \\ \phi(v_j) &= \sum_{k=1}^{n} a_{jk} v_k . \end{aligned}$$

Define the matrix $C = (c_{ij})$ by

$$w_j = \sum_{k=1}^{n} c_{jk} v_k \quad \text{(change-of-basis matrix)},$$

and notice that C is nonsingular by Proposition 3 (show this). We calculate $\phi(w_i)$ two ways:

$$\phi(w_i) = \sum_{j=1}^{n} b_{ij} w_j = \sum_{j=1}^{n} b_{ij} \sum_{k=1}^{n} c_{jk} v_k = \sum_{k=1}^{n} \left(\sum_{j=1}^{n} b_{ij} c_{jk} \right) v_k$$

$$\begin{aligned} \phi(w_i) = \phi\left(\sum_{j=1}^{n} c_{ij} v_j \right) &= \sum_{j=1}^{n} c_{ij} \left(\sum_{k=1}^{n} a_{jk} v_k \right) \\ &= \sum_{k=1}^{n} \left(\sum_{j=1}^{n} c_{ij} a_{jk} \right) v_k . \end{aligned}$$

This shows

$$\begin{aligned} BC &= CA \quad \text{or that} \\ B &= CAC^{-1} . \quad \blacksquare \end{aligned}$$

Suppose f is a function whose domain is the set of $n \times n$ matrices. If f is constant on equivalence classes of similar matrices (i.e., $f(A) = f(CAC^{-1})$ for every nonsingular matrix C), then we get an induced function on $\mathrm{End}(V)$ by Proposition 4. That is, given $\phi \in \mathrm{End}(V)$, we define $f(\phi)$ to be $f(M_\alpha(\phi))$ where α is any basis for V.

An important example of such a function is the determinant function, which we shall study shortly. Another such is the trace function.

Example

We define the *trace* of the $n \times n$ matrix $A = (a_{ij})$ to be the sum of its diagonal elements, i.e.,

$$\text{trace } A = a_{11} + a_{22} + \cdots + a_{nn} .$$

It is left as an exercise to show that similar matrices have the same trace.

<u>Exercises</u>

(1) Verify that similarity is an equivalence relation. (Note that if X and Y are nonsingular, then so is their product XY.)

(2) Given the linear map $\phi : V \to W$ and bases $\alpha = \{v_1, \ldots, v_n\}$ for V and $\beta = \{w_1, \ldots, w_m\}$ for W, we have

$$\phi(v_i) = a_{i1}w_1 + \cdots + a_{im}w_m$$

as before; let

$${}^tM(\phi) = \begin{pmatrix} a_{11} & a_{21} & \cdots & a_{n1} \\ a_{12} & a_{22} & \cdots & a_{n2} \\ \vdots & & & \\ a_{1m} & a_{2m} & \cdots & a_{nm} \end{pmatrix}$$

and note that the ith column of ${}^tM(\phi)$ is the ith row of $M(\phi)$ (see equation (2)). Given the bases α and β, we will see that ${}^tM(\phi)$ is the matrix associated to ϕ when we write vectors as column matrices.

(i) Given $x = x_1v_1 + \cdots + x_nv_n \in V$, we can write this as the $n \times 1$ matrix $\begin{pmatrix} x_1 \\ \vdots \\ x_n \end{pmatrix}$. Similarly, we can write $\phi(x) = y = y_1w_1 + \cdots + y_mw_m$ as $\begin{pmatrix} y_1 \\ \vdots \\ y_m \end{pmatrix}$.

Now show that

$$ {}^tM(\phi)\begin{pmatrix} x_1 \\ \vdots \\ x_n \end{pmatrix} = \begin{pmatrix} y_1 \\ \vdots \\ y_m \end{pmatrix}. $$

(ii) Given the linear maps

$$ V \xrightarrow{\phi} W \xrightarrow{\psi} U , $$

show that we have

$$ {}^tM(\psi \circ \phi) = {}^tM(\psi) \cdot {}^tM(\phi) $$

once we have chosen bases for V, W, and U.

(3) Define $\phi : \mathbf{R}^2 \to \mathbf{R}^2$ by $\phi(x_1, x_2) = (2x_1 - 3x_2, -x_1 - 8x_2)$, and determine $M_\alpha(\phi)$, $M_\beta(\phi)$ where α is the standard basis and $\beta = \{v_1 = (1,1), v_2 = (-1,2)\}$. Find C such that $M_\beta(\phi) = CM_\alpha(\phi)C^{-1}$.

The Rank of a Matrix.

Given the $n \times m$ matrix $A = (a_{ij})$ with entries form the field k, we denote the rows of A by $A_1, \ldots, A_n$ and the columns of A by $A^1, \ldots, A^m$, i.e., A_i is the vector $(a_{i1}, \ldots, a_{im}) \in k^m$ and A^j is the vector ${}^t(a_{1j}, \ldots, a_{nj}) \in k^n$.

Definition The *row space* of A is the subspace of k^m generated by $\{A_1, \ldots, A_n\}$, and the *column space* of A is the subspace of k^n generated by $\{A^1, \ldots, A^m\}$. The row (column) rank of A is the dimension of the row (column) space of A.

An interesting fact is that the row and column ranks of A are equal. Thus we simply refer to the *rank* of A. It is also true that similar matrices have the same rank, so we define the *rank* of $\phi \in \mathrm{End}(V)$ to be the rank of any matrix which represents ϕ.

More Exercises

(4) Show that the row rank and column rank of any matrix A are equal. (Hint: Choose a maximal linearly independent subset $\{B_1, \ldots, B_r\}$ of $\{A_1, \ldots, A_n\}$ and write

$$ (*) \begin{cases} A_1 &= c_{11}B_1 + \cdots + c_{1r}B_r \\ \vdots & \\ A_n &= c_{n1}B_1 + \cdots + c_{nr}B_r \end{cases} $$

for scalars $c_{ij} \in k$. Now note that the jth components on the left hand side of $(*)$ combine to give A^j.)

(5) Show that if the square matrices A, B are similar, then they have the same rank. (Hint: Think of the $n \times n$ matrix A as an element of $\mathrm{End}(k^n)$ and observe that the rank of A is just the dimension of im (A).)

(6) What is the rank of the following matrices?

$$\text{(i) } A = \begin{pmatrix} 1 & 0 \\ -1 & 1 \\ 2 & 3 \end{pmatrix}$$

$$\text{(ii) } A = \begin{pmatrix} 1 & 0 & -1 \\ 2 & 1 & 0 \\ 1 & -1 & -3 \end{pmatrix}$$

(7) Given the matrix A, if we delete some (or possibly none) of its rows and/or columns, then we obtain a *submatrix* of A. For example,

$$A' = \begin{pmatrix} 1 & 0 \\ 2 & 3 \end{pmatrix}$$

is the square submatrix that we get from A in exercise (6i) when we delete its second row. Show that r is the rank of the matrix $A \Leftrightarrow A$ has an $r \times r$ square submatrix which is nonsingular, but any larger square submatrix of A is singular. (Note that this will be easy to prove once we have developed the theory of elementary row operations.)

(8) Given $n \times n$ matrices A and B, show that $\mathrm{trace}(AB) = \mathrm{trace}(BA)$. Now show that similar matrices have the same trace.

B. Algebras

Let V be a vector space over a field k. If V also has a multiplication on it which behaves well with respect to the vector space operations, then V is called an *algebra* (see Section B of Chapter I). Precisely, the multiplication must distribute over vector addition

$$v(w+u) = vw + vu \quad \text{and} \quad (w+u)v = wv + uv \tag{5}$$

and for scalar multiplication we need

$$v(rw) = r(vw) = (rv)w \qquad \text{for} \quad r \in k. \tag{6}$$

Note that the zero vector 0 of V is a "multiplicative annihilator" because

$$0v = (0+0)v = 0v + 0v \quad .$$

and, since V is an additive group, this implies $0v = 0$.

The algebra V is *associative* if we always have $v(wu) = (vw)u$. It is *commutative* if we always have $vw = wv$. The algebra has a *unit element* if there is a $1 \in V$ such that $1v = v1 = v$ for each $v \in V$. (Note that such an element must be unique. See Proposition 1 of Chapter 0.)

Example

Let $M_n(k)$ denote the set of all $n \times n$ matrices over the field k. We first make $M_n(k)$ into a vector space by defining:

for $A = (a_{ij})$ and $B = (b_{ij})$ in $M_n(k)$, let

$$A + B = (a_{ij} + b_{ij}), \qquad \text{and}$$

for $A = (a_{ij})$ and $r \in k$, we set

$$rA = (ra_{ij}).$$

A little thought will make it clear that this is just the vector space k^{n^2} of all ordered n^2-tuples of elements of k. We are just writing the n^2-tuples as square arrays.

In the previous section we saw how to multiply $n \times n$ matrices and we readily verify

$$\begin{aligned} A(B+C) &= AB + AC \\ (B+C)A &= BA + CA \\ A(rB) &= r(AB) = (rA)B, \end{aligned}$$

so that $M_n(k)$ becomes an algebra. Also it is associative since matrix multiplication corresponds to composition of linear maps which is manifestly associative. But $M_n(k)$ is not commutative (for $n > 1$). For example,

$$\begin{pmatrix} 1 & 0 \\ 1 & 0 \end{pmatrix}\begin{pmatrix} 0 & 0 \\ 1 & 0 \end{pmatrix} = \begin{pmatrix} 0 & 0 \\ 0 & 0 \end{pmatrix}$$

whereas

$$\begin{pmatrix} 0 & 0 \\ 1 & 0 \end{pmatrix}\begin{pmatrix} 1 & 0 \\ 1 & 0 \end{pmatrix} = \begin{pmatrix} 0 & 0 \\ 1 & 0 \end{pmatrix}.$$

This algebra has a unit element, namely, the identity matrix

$$I = \begin{pmatrix} 1 & \dots & & 0 \\ & \ddots & & \\ \vdots & & 1 & \vdots \\ & & & \ddots \\ 0 & & \dots & 1 \end{pmatrix} = (\delta_{ij}), \qquad \delta_{ij} = \begin{cases} 1 & \text{if } i = j \\ 0 & \text{if } i \neq j. \end{cases}$$

This clearly satisfies $IA = AI = A$ for any $A \in M_n(k)$. The algebras $M_n(k)$ are *matrix algebras* over k.

We have seen that once we choose a basis $\alpha = \{v_1, \ldots, v_n\}$ for the n-dimensional vector space V over the field k, we get an isomorphism

$$V \cong k^n$$

which takes $x = x_1v_1 + \cdots + x_nv_n \in V$ to its coordinate vector $(x_1, \ldots, x_n) \in k^n$. This choice of basis also gives a one-to-one correspondence between endomorphisms $\phi \in \mathrm{End}(V)$ and $n \times n$ matrices $M_\alpha(\phi) \in M_n(k)$; this correspondence is an (algebra) isomorphism by exercise (2) of section A. Unfortunately, this isomosphism is not "natural" since it depends on α. (By Proposition 3, we know how a different choice of basis affects this isomorphism.)

Henceforth, we will sometimes use ϕ and $M_\alpha(\phi)$ interchangeably (especially when $V = k^n$ and α is the standard basis). We have already observed the potential difficulty in doing this (see the "important observation" in section A).

Example

We have already made the plane $\mathbf{R}^2$ into a vector space. Now we make it into an algebra by defining

$$(a, b)(c, d) = (ac, bd). \tag{7}$$

It is easy to verify that (7) does make $\mathbf{R}^2$ into an algebra. Also this algebra is associative and commutative and it has unit element $(1, 1)$. But (7) does not make $\mathbf{R}^2$ into a field, because we have seen that fields have no divisors of zero whereas (7) implies $(1, 0)(0, 1) = (0, 0)$.

We saw in Chapter 0 that a "better" multiplication exists on $\mathbf{R}^2$; namely

$$(a, b)(c, d) = (ac - bd, ad + bc). \tag{8}$$

This is associative and commutative, has unit element $(1, 0)$, and is a field.

Example: The exterior algebra on $\mathbf{R}^3$

We will introduce here the *exterior algebra* Λ on $\mathbf{R}^3$. (In the next chapter we will define it for all $\mathbf{R}^n$.)

As a vector space, Λ is going to be the direct sum of four vector spaces

$$\Lambda = \Lambda^0 \oplus \Lambda^1 \oplus \Lambda^2 \oplus \Lambda^3. \tag{9}$$

Λ^0 is just the vector space $\mathbf{R}$ with basis $\{1\}$. Λ^1 is the vector space $\mathbf{R}^3$ with the standard basis $\{e_1, e_2, e_3\}$.

Now Λ^2 is going to be another three-dimensional real vector space and we are going to write the basis elements as

$$\{e_1 \wedge e_2, e_1 \wedge e_3, e_2 \wedge e_3\}.$$

Finally, Λ^3 is going to be a one-dimensional real vector space with basis element $\{e_1 \wedge e_2 \wedge e_3\}$.

Our clumsy notation for basis elements (at least for Λ^2 and Λ^3) is in preparation for making the vector space

$$\Lambda = \Lambda^0 \oplus \Lambda^1 \oplus \Lambda^2 \oplus \Lambda^3$$

into an algebra. Now Λ is an eight-dimensional real vector space with basis

$$\{1, e_1, e_2, e_3, e_1 \wedge e_2, e_1 \wedge e_3, e_2 \wedge e_3, e_1 \wedge e_2 \wedge e_3\} \tag{10}$$

consisting of "four kinds" of elements.

We are going to define a multiplication on Λ which will distribute over vector addition and thus it will suffice to define multiplication of basis elements. Our algebra is also going to be associative.

(i) First we decree that the basis element 1 for $\Lambda^0 (= \mathbf{R})$ is to be the unit element for the multiplication.

(ii) Next we decide that the product of two elements in Λ^1 is to be in Λ^2. Precisely, we take:

the product of e_1 and e_2 is to be $e_1 \wedge e_2$,
the product of e_1 and e_3 is to be $e_1 \wedge e_3$, and
the product of e_2 and e_3 is to be $e_2 \wedge e_3$.

(iii) We also need to know the product of e_2 and e_1, e_1 and e_1, etc. We set

$$e_j \wedge e_i = -(e_i \wedge e_j) \tag{11}$$

for all i, j — even for $i = j$.

In particular, $e_i \wedge e_i = -(e_i \wedge e_i)$ and this must be the zero vector (in Λ^2) since $2 \neq 0$ in $\mathbf{R}$.

So the multiplication of basis elements in Λ^1 is given by

	e_1	e_2	e_3
e_1	0	$e_1 \wedge e_2$	$e_1 \wedge e_3$
e_2	$-(e_1 \wedge e_2)$	0	$e_2 \wedge e_3$
e_3	$-(e_1 \wedge e_3)$	$-(e_2 \wedge e_3)$	0

and using distributivity we can calculate the product of any two elements of Λ^1.

The general rule (11) also gives us guidance in determining other products. For example, suppose we want to multiply $e_3 \in \Lambda^1$ with $e_1 \wedge e_2 \in \Lambda^2$. We write the product as $e_3 \wedge (e_1 \wedge e_2)$, and we are insisting on associativity, so we can write this as $e_3 \wedge e_1 \wedge e_2$. By (11), this equals

$$-e_1 \wedge e_3 \wedge e_2 = e_1 \wedge e_2 \wedge e_3.$$

As another example consider $(e_1 \wedge e_2) \wedge (e_2 \wedge e_3)$. This can be written $e_1 \wedge (e_2 \wedge e_2) \wedge e_3 = 0$, since $e_2 \wedge e_2 = 0$.

Finally, let's do a fairly general example:

$$\begin{aligned}
&(a_0 + a_1 e_1 + a_3 e_3 + a_{13} e_1 \wedge e_3 + a_{123} e_1 \wedge e_2 \wedge e_3) \\
&\wedge (b_2 e_2 + b_{12} e_1 \wedge e_2 + b_{13} e_1 \wedge e_3) \\
&= a_0 b_2 e_2 + a_0 b_{12} e_1 \wedge e_2 + a_0 b_{13} e_1 \wedge e_3 \\
&\quad + a_1 b_2 e_1 \wedge e_2 + 0 + 0 \\
&\quad + a_3 b_2 e_3 \wedge e_2 + a_3 b_{12} e_3 \wedge e_1 \wedge e_2 + 0 \\
&\quad + a_{13} b_2 e_1 \wedge e_3 \wedge e_2 + 0 + 0 \\
&\quad + 0 + 0 + 0 \\
&= a_0 b_2 e_2 + (a_0 b_{12} + a_1 b_2) e_1 \wedge e_2 + a_0 b_{13} e_1 \wedge e_3 \\
&\quad - a_3 b_2 e_2 \wedge e_3 + (a_3 b_{12} - a_{13} b_2) e_1 \wedge e_2 \wedge e_3.
\end{aligned}$$

You are asked to do a few more in the exercises.

Note that the product of an element in Λ^p and an element in Λ^q will be an element in Λ^{p+q}. (Of course $\Lambda^{p+q} = \{0\}$ for $p + q > 3$.)

Note that in our definition of algebra we did not require the multiplication to be associative or commutative. Given elements x, y, z of an algebra $\mathcal{A}$, we define their *associator* by

$$[x, y, z] = (xy)z - x(yz). \tag{12}$$

Clearly $\mathcal{A}$ is associative $\Leftrightarrow$ all associators are zero. For two elements x, y of $\mathcal{A}$ we define their *commutator* by

$$[x, y] = xy - yx. \tag{13}$$

$\mathcal{A}$ is commutative $\Leftrightarrow$ all commutators are zero.

Recall that the algebra $\text{End}(V)$ is associative but not commutative. From $\text{End}(V)$ we construct a new kind of algebra, a *Lie Algebra*, by defining the product of $x, y \in \text{End}(V)$ to be their commutator

$$[x, y] = xy - yx.$$

Clearly

$$[y, x] = -[x, y]. \tag{14}$$

So, in particular, $[x, x] = 0$ for all $x \in \text{End}(V)$.

This algebra fails to be associative, but satisfies the *Jacobi identity*

$$[x, [y, z]\,] + [y, [z, x]\,] + [z, [x, y]\,] = 0, \tag{15}$$

the proof of which is a routine exercise.

Theorem 1 *Let $\mathcal{A}$ be an associative algebra. Then $\mathcal{A}$ with the new multiplication*

$$[x, y] = xy - yx$$

is a Lie Algebra (i.e., $[x, y] = -[y, x]$ and the Jacobi identity is satisfied).

Example

There is an important three-dimensional Lie Algebra. Let $x = (1,0,0)$, $y = (0,1,0)$, $h = (0,0,1)$ in $\mathbf{R}^3$. Define

$$[h,x] = 2x, \quad [h,y] = -2y, \quad [x,y] = h, \quad \text{etc.}$$

(This is the Lie Algebra associated with the Lie group whose elements are quaternians of unit length.)

Exercises

(1) Compute the following products:

(i) $(e_2 \wedge e_1) \wedge (e_2 \wedge e_3)$ (Note that this can be done two different ways: either using (11) or the fact that $\Lambda^4 = \{0\}$.)

(ii) $(e_1 + e_2 \wedge e_3) \wedge (e_1 + e_2 \wedge e_3)$

(iii) $(3e_2 - 2e_1 \wedge e_2 + 4e_2 \wedge e_3) \wedge (2 - 5e_1)$

(2) From (11), we concluded that $e_i \wedge e_i = 0$. Given any $w \in \Lambda$, can we conclude that $w \wedge w = 0$? (Hint: See (ii) above.)

(3) Prove Theorem 1, i.e., show that the new multiplication satisfies (5), (6), and (15).

C. Determinants, the Laplace Expansion

We will now define the exterior algebra Λ "over" $\mathbf{R}^n$ (for $n = 3$, see Section B). We fix n. As a vector space, Λ will be a direct sum

$$(16) \qquad \Lambda = \Lambda^0 \oplus \Lambda^1 \oplus \ldots \oplus \Lambda^n.$$

Just as before, $\Lambda^0 = \mathbf{R}$ with basis $\{1\}$. Similarly, $\Lambda^1 = \mathbf{R}^n$ with basis $\{e_1, e_2, \ldots, e_n\}$. Then Λ^2 will be a real vector space with basis $\{e_i \wedge e_j | 1 \leq i < j \leq n\}$. Λ^3 will have basis $\{e_i \wedge e_j \wedge e_k | 1 \leq i < j < k \leq n\}$, etc. Finally, Λ^n will have basis $\{e_1 \wedge e_2 \wedge \ldots \wedge e_n\}$.

We note that the dimension of Λ^p $(p = 0, 1, \ldots, n)$ is $\dfrac{n!}{p!(n-p)!}$.

Exercise. Show that $\dim(\Lambda) = 2^n$.

For the algebra structure we insist (just as we did for $n = 3$) that:

(i) multiplication is to be associative,

(ii) multiplication is to distribute over addition, and

(iii) for all i, j, $e_i \wedge e_j = -(e_j \wedge e_i)$.

Just as for $n = 3$, we see that these conditions suffice to define the multiplication. If $x \in \Lambda^p$ and $y \in \Lambda^q$, then $x \wedge y \in \Lambda^{p+q}$ (of course if $p+q > n$, then $\Lambda^{p+q} = \{0\}$).

If in $e_{i_1} \wedge e_{i_2} \wedge \ldots \wedge e_{i_m}$ any two are equal, then this product is zero. For by (iii), we can move the two equal ones together by sign changes and then (again by (iii)) the product is zero.

Proposition 5 *If $x, y \in \Lambda^1$ are linearly dependent, then $x \wedge y = 0$.*

Proof: For any $x, y \in \Lambda^1$ we can write $x = x_1e_1 + \cdots + x_ne_n$, $y = y_1e_1 + \cdots + y_ne_n$ and, using (i), (ii), and (iii), we have

$$x \wedge y = \sum_{i,j} x_iy_je_i \wedge e_j = \sum_{i<j}(x_iy_j - x_jy_i)e_i \wedge e_j.$$

Now suppose that for some $r \in \mathbf{R}$, $y = rx$. Then each $y_i = rx_i$ and

$$x \wedge y = \sum_{i<j}(x_irx_j - x_jrx_i)e_i \wedge e_j = 0. \quad \blacksquare$$

Corollary For $x \in \Lambda^1$, $x \wedge x = 0$.

Proposition 6 *If $v_1, v_2, \ldots, v_m \in \Lambda^1$ are linearly dependent, then $v_1 \wedge v_2 \wedge \ldots \wedge v_m = 0$.*

Proof: We have $0 = a_1v_1 + \cdots + a_mv_m$ with not all a_i zero. We may as well say $a_1 \neq 0$. Then

$$v_1 = b_2v_2 + \cdots + b_mv_m \text{ (with } b_2 = -\frac{a_2}{a_1}, \text{ etc.)}.$$

So we have

$$\begin{aligned} v_1 \wedge v_2 \wedge \cdots \wedge v_m = \; & b_2v_2 \wedge v_2 \wedge v_3 \wedge \cdots \wedge v_m \\ & + b_3v_3 \wedge v_2 \wedge v_3 \wedge \cdots \wedge v_m \\ & \vdots \\ & + b_mv_m \wedge v_2 \wedge v_3 \wedge \cdots \wedge v_m \,. \end{aligned}$$

Each term has a repeated v_j and is thus zero. $\blacksquare$

It is important to notice that the contrapositive of Proposition 6 is a *criterion for linear independence*: if $v_1, \ldots, v_m \in \mathbf{R}^n (= \Lambda^1)$ satisfy $v_1 \wedge \cdots \wedge v_m \neq 0$, then $v_1, \ldots, v_m$ are linearly independent.

We are now ready to discuss determinants of matrices. In $\mathbf{R}^n$, $e_1, \ldots, e_n$ are ordered n-tuples and can be considered to be $1 \times n$ matrices (or "row vectors"). So given an $n \times n$ real matrix A, the matrix product

$$e_i A$$

is again a $1 \times n$ matrix which we see is just the ith row of the matrix A. Thus each $e_i A \in \mathbf{R}^n = \Lambda^1$. It follows that

$$e_1 A \wedge \cdots \wedge e_n A$$

lies in the one-dimensional vector space Λ^n and thus is some real multiple of the standard basis vector $e_1 \wedge e_2 \wedge \cdots \wedge e_n$ for Λ^n, i.e.,

$$e_1 A \wedge \cdots \wedge e_n A = (\det A) e_1 \wedge \cdots \wedge e_n. \tag{17}$$

Definition The real number $\det A$ is called the *determinant* of the real $n \times n$ matrix A.

Example

We calculate $\det A$ for $A = \begin{pmatrix} a & b \\ c & d \end{pmatrix}$

$$\begin{aligned} e_1 A &= a e_1 + b e_2, \quad e_2 A = c e_1 + d e_2 \\ e_1 A \wedge e_2 A &= (a e_1 + b e_2) \wedge (c e_1 + d e_2) \\ &= ad\, e_1 \wedge e_2 + bc\, e_2 \wedge e_1 \\ &= (ad - bc) e_1 \wedge e_2. \end{aligned}$$

So $\det \begin{pmatrix} a & b \\ c & d \end{pmatrix} = ad - bc.$

Notice that (17) not only defines $\det A$, but (using (i), (ii), and (iii)) also furnishes a means for calculating $\det A$.

Proposition 7

(i) *If I is the $n \times n$ identity matrix, then*

$$\det I = 1\,.$$

(ii) *If the rows of A are linearly dependent, then*

$$\det A = 0\,.$$

(iii) *For $r \in \mathbf{R}$, $\det(rA) = r^n(\det A)$.*

Proof:

(i) The rows of I are $e_1, \ldots, e_n$ so

$$e_1 \wedge \cdots \wedge e_n = (\det I) e_1 \wedge \cdots \wedge e_n.$$

(ii) By Proposition 6, $e_1 A \wedge \cdots \wedge e_n A = 0$.

(iii) The rows of rA are $re_1 A, \ldots, re_n A$ and

$$re_1 A \wedge \cdots \wedge re_n A = r^n (e_1 A \wedge \cdots \wedge e_n A) = r^n (\det A) e_1 \wedge \cdots \wedge e_n. \quad ∎$$

Let C be an $n \times n$ real matrix. Then we know how C operates on Λ^1 and we now define an operation of C on a product $v_1 \wedge \cdots \wedge v_p \quad (v_i \in \Lambda^1)$ by

$$(v_1 \wedge \cdots \wedge v_p) C = v_1 C \wedge \cdots \wedge v_p C . \tag{18}$$

Now we get a very important property of

$$\det : M_n(\mathbf{R}) \to \mathbf{R}.$$

Proposition 8 *Let A, B be $n \times n$ real matrices. Then*

$$\det(AB) = (\det A)(\det B). \tag{19}$$

Proof: By definition

$$e_1(AB) \wedge \cdots \wedge e_n(AB) = \det(AB) e_1 \wedge \cdots \wedge e_n.$$

Also

$$\begin{aligned} e_1(AB) \wedge \cdots \wedge e_n(AB) &= (e_1 A \wedge \cdots \wedge e_n A) B \quad \text{(by (18))} \\ &= (\det A)(e_1 \wedge \cdots \wedge e_n) B \quad \text{(by (17))} \\ &= (\det A)(e_1 B \wedge \cdots \wedge e_n B) = (\det A)(\det B) e_1 \wedge \cdots \wedge e_n . \quad ∎ \end{aligned}$$

A very useful property of det is:

Proposition 9

$$A \textit{ is nonsingular} \Leftrightarrow \det A \neq 0. \tag{20}$$

Proof: $\Rightarrow$ By hypothesis A^{-1} exists. So $1 = \det I = \det(AA^{-1}) = (\det A)(\det A^{-1})$, proving $\det A \neq 0$.

$\Leftarrow$ By (ii) of Proposition 7, the rows of A (which are the images of $e_1, \ldots, e_n$ under the linear map A) are linearly independent. Thus A is an isomorphism of $\mathbf{R}^n$ and A^{-1} exists. ∎

Recall that two $n \times n$ matrices A, B are similar if there is a nonsingular $n \times n$ matrix C such that $B = CAC^{-1}$.

Proposition 10 det : $M_n(\mathbf{R}) \to \mathbf{R}$ *is constant on equivalence classes of similar matrices.*

Proof: If $B = CAC^{-1}$, then $\det(B) = \det(CAC^{-1}) = \det(C)\det(A)\det(C^{-1})$ and $(\det C)(\det C^{-1}) = \det(CC^{-1}) = \det I = 1$. Thus

$$\det B = \det A . \quad \blacksquare$$

Observation:

Notice that our development of the determinant function has only used the fact that $\mathbf{R}$ is a field — we have not used any properties unique to the real numbers. Thus, for any field k, we get a determinant function

$$\det : M_n(k) \to k$$

which is constant on equivalence classes of similar matrices, etc.

Exercises

(1) A matrix $D = (d_{ij}) \in M_n(k)$ is *diagonal* if $d_{ij} = 0$ whenever $i \neq j$. Show that if D is diagonal, then $\det D = d_{11}d_{22}\ldots d_{nn}$.

(2) A matrix $A = (a_{ij})$ is *(weakly) upper triangular* if $a_{ij} = 0$ whenever $i > j$. Show that this implies $\det A = a_{11}a_{22}\ldots a_{nn}$. A is *lower triangular* if $a_{ij} = 0$ whenever $i < j$. Show that this also implies $\det A = a_{11}\ldots a_{nn}$. ($A$ is called *triangular* if it is either upper or lower triangular.)

(3) The *transpose* ${}^tA$ of matrix $A = (a_{ij})$ is ${}^tA = (a_{ji})$. (That is, the rows of ${}^tA$ are the columns of A.) Show that ${}^t({}^tA) = A$ and that $\det {}^tA = \det A$.

(4) Define the (real) *general linear group* to be all nonsingular matrices $A \in M_n(\mathbf{R})$; i.e.,

$$GL(n, \mathbf{R}) = \{A \in M_n(\mathbf{R}) \mid \det A \neq 0\} .$$

Show that $GL(n, \mathbf{R})$ is, indeed, a group. Note that $GL(n, \mathbf{R})$ is simply $GL(\mathbf{R}^n)$.

(5) Let

$$A = \begin{pmatrix} a_{11} & a_{12} & a_{13} \\ a_{21} & a_{22} & a_{23} \\ a_{31} & a_{32} & a_{33} \end{pmatrix} .$$

Show that

$$\det A = a_{11}\det\begin{pmatrix} a_{22} & a_{23} \\ a_{32} & a_{33} \end{pmatrix} - a_{12}\det\begin{pmatrix} a_{21} & a_{23} \\ a_{31} & a_{33} \end{pmatrix} + a_{13}\det\begin{pmatrix} a_{21} & a_{22} \\ a_{31} & a_{32} \end{pmatrix}.$$

(This is a special case of the "Laplace expansion.")

The Laplace Expansion

Recall that our definition of $\det A$ for an $n \times n$ matrix A is

$$e_1 A \wedge \cdots \wedge e_n A = (\det A) e_1 \wedge \cdots \wedge e_n \,. \tag{21}$$

We can rearrange things in this product to get some formulas for $\det A$ in terms of some determinants of $(n-1)\times(n-1)$ matrices. This can be useful sometimes in calculating $\det A$, and can be very useful in getting inductive proofs of certain theorems about determinants.

First off, let us rewrite (21) as

$$(a_{11}e_1 + \cdots + a_{1n}e_n) \wedge \cdots \wedge (a_{n1}e_1 + \cdots + a_{nn}e_n) = (\det A) e_1 \wedge \cdots \wedge e_n$$

Next we do the case $n = 3$ to make the procedure clear without too much notation.

$$\left.\begin{aligned}
&(a_{11}e_1 + a_{12}e_2 + a_{13}e_3) \wedge (a_{21}e_1 + a_{22}e_2 + a_{23}e_3) \wedge (a_{31}e_1 + a_{32}e_2 + a_{33}e_3) \\
&= a_{11}e_1 \wedge (a_{21}e_1 + a_{22}e_2 + a_{23}e_3) \wedge (a_{31}e_1 + a_{32}e_2 + a_{33}e_3) \\
&\quad + a_{12}e_2 \wedge (a_{21}e_1 + a_{22}e_2 + a_{23}e_3) \wedge (a_{31}e_1 + a_{32}e_2 + a_{33}e_3) \\
&\quad + a_{13}e_3 \wedge (a_{21}e_1 + a_{22}e_2 + a_{23}e_3) \wedge (a_{31}e_1 + a_{32}e_2 + a_{33}e_3) \\
&= a_{11}e_1 \wedge (a_{22}e_2 + a_{23}e_3) \wedge (a_{32}e_2 + a_{33}e_3) \\
&\quad + a_{12}e_2 \wedge (a_{21}e_1 + a_{23}e_3) \wedge (a_{31}e_1 + a_{33}e_3) \\
&\quad + a_{13}e_3 \wedge (a_{21}e_1 + a_{22}e_2) \wedge (a_{31}e_1 + a_{32}e_2) \\
&= a_{11}a_{22}a_{33}\; e_1 \wedge e_2 \wedge e_3 + a_{11}a_{23}a_{32}\; e_1 \wedge e_3 \wedge e_2 \\
&\quad + a_{12}a_{21}a_{33}\; e_2 \wedge e_1 \wedge e_3 + a_{12}a_{23}a_{31}\; e_2 \wedge e_3 \wedge e_1 \\
&\quad + a_{13}a_{21}a_{32}\; e_3 \wedge e_1 \wedge e_2 + a_{13}a_{22}a_{31}\; e_3 \wedge e_2 \wedge e_1 \\
&= \{a_{11}(a_{22}a_{33} - a_{23}a_{32}) \\
&\quad - a_{12}(a_{21}a_{33} - a_{23}a_{31}) \\
&\quad + a_{13}(a_{21}a_{32} - a_{22}a_{31})\}\; e_1 \wedge e_2 \wedge e_3
\end{aligned}\right\} \quad (*)$$

Let B_{ij} be the 2×2 matrix obtained from A by deleting the ith row and j^{th} column. Then we have proved that

$$\det A = a_{11} \det B_{11} - a_{12} \det B_{12} + a_{13} \det B_{13} .$$

For an $n \times n$ matrix A, the definition of B_{ij} (an $(n-1) \times (n-1)$ matrix) is similar.

Proposition 11 *For an $n \times n$ matrix A, we have*

$$(22) \quad \det A = a_{11} \det B_{11} - a_{12} \det B_{12} + \cdots + (-1)^{n+1} a_{1n} \det B_{1n}.$$

Proof: Same as for $n = 3$. ∎

Equation (22) is the *Laplace expansion* about the first row of A. More generally, we can calculate $\det A$ by computing the Laplace expansion about the ith row of A, i.e.,

$$\det A = (-1)^{i+1} a_{i1} \det B_{i1} + \cdots + (-1)^{i+n} a_{in} \det B_{in} .$$

Proposition 12 *For an $n \times n$ matrix A we have*

$$(23) \quad \det A = a_{11} \det B_{11} - a_{21} \det B_{21} + \cdots + (-1)^{n+1} a_{n1} \det B_{n1}.$$

(This is the Laplace expansion about the first column of A. We can also calculate $\det A$ using the Laplace expansion about the jth column of A. Exercise. Write out the appropriate formula.)

Proof: Again the general case should be clear from the computation for $n = 3$.

We rewrite (*) as

$$\begin{aligned}
&(a_{11}e_1 \wedge (a_{22}e_2 + a_{23}e_3) \wedge (a_{32}e_2 + a_{33}e_3) \\
&+ (a_{12}e_2 + a_{13}e_3) \wedge (a_{21}e_1) \wedge (a_{32}e_2 + a_{33}e_3) \\
&+ (a_{12}e_2 + a_{13}e_3) \wedge (a_{22}e_2 + a_{23}e_3) \wedge (a_{31}e_1)
\end{aligned}$$

and this equals

$$\begin{aligned}
&a_{11}a_{22}a_{33}\, e_1 \wedge e_2 \wedge e_3 + a_{11}a_{23}a_{32}\, e_1 \wedge e_3 \wedge e_2 \\
&+ a_{21}a_{12}a_{33}\, e_2 \wedge e_1 \wedge e_3 + a_{21}a_{13}a_{32}\, e_3 \wedge e_1 \wedge e_2 \\
&+ a_{31}a_{12}a_{23}\, e_2 \wedge e_3 \wedge e_1 + a_{31}a_{13}a_{22}\, e_3 \wedge e_2 \wedge e_1 \\
&= \{a_{11}(a_{22}a_{33} - a_{23}a_{32}) - a_{21}(a_{12}a_{33} - a_{13}a_{32}) \\
&\quad + a_{31}(a_{12}a_{23} - a_{13}a_{22})\}\, e_1 \wedge e_2 \wedge e_3 .
\end{aligned}$$

Thus

$$\det A = a_{11} \det B_{11} - a_{21} \det B_{21} + a_{31} \det B_{31}. \quad ∎$$

Exercises

(1) Recall that we define the transpose ${}^tA = ({}^ta_{ij})$ of an $n \times n$ matrix $A = (a_{ij})$ by ${}^ta_{ij} = a_{ji}$. (Definition: A is *symmetric* if $A = {}^tA$.) Prove the following.

(i) ${}^t({}^tA) = A$, ${}^t(A+B) = {}^tA + {}^tB$.

(ii) $A + {}^tA$ is symmetric.

(iii) ${}^t(AB) = {}^tB\,{}^tA$.

(iv) $A\,{}^tA$ is symmetric.

(v) $\det {}^tA = \det A$.
(Hint: Use the Laplace expansion about the first row of A and the first column of ${}^tA$. Then use induction.)

(vi) If the nonsingular matrix A is symmetric, then A^{-1} is also symmetric.

(2) Show that the determinant of $\begin{pmatrix} A & C \\ 0 & B \end{pmatrix}$ is $(\det A)(\det B)$, where A and B and square submatrices. Generalize this to calculating the determinant of

$$\begin{pmatrix} A_1 & & & \\ & A_2 & & * \\ & & \ddots & \\ 0 & & & A_m \end{pmatrix}$$

where $A_1, \ldots, A_m$ are square submatrices.

Elementary Row Operations.

(3) Let $A = (a_{ij})$ be an element of $M_n(k)$. The following three operations are called *elementary row operations.*

E_1 : Interchange the ith row and j^{th} row
E_2 : Multiply a row by a nonzero scalar $r \in k$
E_3' : Replace i^{th} row by (i^{th} row) + (j^{th} row)

Two matrices M, N are *row equivalent* if N can be obtained from M by a finite sequence of elementary row operations. Show that this is an equivalence relation.

(4) Let E_3 be the operation: E_2 and then E_3'. Show that E_1, E_2, E_3 still give the same equivalence relation.

(5) Show that row equivalent matrices have the same row space.

(6) An $n \times n$ matrix $A = (a_{ij})$ is in *echelon form* if the number of zeros preceding the first nonzero entry in a row increases row by row, stopping this increase only when arriving at an all zero row (then succeeding rows must be all zeros). The first nonzero elements in rows of an echelon matrix are called *distinguished elements.* Prove that any matrix can be reduced to echelon form by an appropriate sequence of elementary row operations E_1, E_2, E_3.

(7) Show that r is the rank of the $n \times n$ matrix $A \Leftrightarrow A$ has a nonsingular $r \times r$ submatrix, but any larger square submatrix of A is singular.

(8) An echelon matrix A is *row reduced* if its distinguished elements satisfy:

(i) each is the only nonzero element in its column, and

(ii) each is equal to 1.

Prove that any echelon matrix can be put in row reduced form by an appropriate sequence of E_1, E_2, E_3.

(9) Let A be an $n \times n$ matrix. Show that $\det A$ is changed by E_1, E_2, and E_3 in the following manner:

(i) $\det(E_1(A)) = -\det A$,

(ii) $\det(E_2(A)) = r(\det A)$, and

(iii) $\det(E_3(A)) = \det A$.

(10) An $n \times n$ matrix e is an *elementary matrix* if e can be obtained from the $n \times n$ identity matrix I by a single $E_i \in \{E_1, E_2, E_3\}$. Show that we can perform an elementary row operation on a matrix by multiplying by the appropriate elementary matrix. Now show that if A is an invertible (= nonsingular) $n \times n$ matrix and A is row reduced to the echelon matrix B by elementary row operations (i.e., there are elementary $n \times n$ matrices $e_1, e_2, \ldots, e_q$ such that $B = e_q \ldots e_2 e_1 A$), then we must have $B = I$. Conclude that invertible $n \times n$ matrices are precisely those which are products of elementary matrices.

Given $A \in M_n(k)$, note that E_1, E_2, E_3 make evaluating $\det A$ easier, since we can simplify A to the matrix B with E_1, E_2, E_3 (being careful to keep track of how these changes relate $\det A$ to $\det B$): we can either row reduce A in which case B is a diagonal matrix (and $\det B$ is simply the product of its diagonal elements); or we can make all (but possibly one) elements in a column of B equal to 0, and then calculate $\det B$ by using the Laplace expansion about this column.

Example

If $A = \begin{pmatrix} 2 & 0 & 1 \\ 1 & 2 & 5 \\ -1 & 4 & 2 \end{pmatrix}$, then subtract twice the second row from the third row to obtain

$$B = \begin{pmatrix} 2 & 0 & 1 \\ 1 & 2 & 5 \\ -3 & 0 & -8 \end{pmatrix}.$$

We have applied the operation E_3 to obtain B, so $\det A = \det B$. We can calculate $\det B$ by expanding about the second column:

$$\det B = 2 \det \begin{pmatrix} 2 & 1 \\ -3 & -8 \end{pmatrix} = 2(-16 - (-3)) = -26\,.$$

More Exercises

(11) Compute $\det A$ for the following matrices.

$$\text{(i) } A = \begin{pmatrix} 2 & 4 & -1 \\ 1 & 3 & 0 \\ 0 & 2 & 1 \end{pmatrix}$$

$$\text{(ii) } A = \begin{pmatrix} 0 & -3 & 1 \\ 0 & 2 & 4 \\ 1 & 0 & -2 \end{pmatrix}$$

$$\text{(iii) } A = \begin{pmatrix} -3 & 0 & 2 & 0 \\ 1 & 2 & -4 & 1 \\ 0 & 3 & -1 & -1 \\ 2 & -2 & 3 & 5 \end{pmatrix}$$

(12) If we replace "row" by "column" in exercise (3), we get the *elementary column operations* on $A \in M_n(k)$. Give appropriate versions of exercises (3)–(10) where the notions of row equivalence, etc., have been replaced by the notions of column equivalence, etc. Now prove these revised exercises. (Hint for (9): recall that $\det {}^tA = \det A$.)

(13) Some of the exercises (3)–(10) generalize to the case of $n \times m$ matrices. Decide which of these exercise can be generalized, give appropriate versions, and then prove them.

D. Inverses, Systems of Equations

For $A = (a_{ij}) \in M_n(k)$, we know that A has an inverse $A^{-1} \Leftrightarrow \det A \neq 0$. But this does not tell us how to calculate A^{-1} when we know it exists. In this section we determine a way to calculate A^{-1}.

For the i,j entry a_{ij} of A, we define its *cofactor* to be

$$b_{ij} = (-1)^{i+j} \det B_{ij}$$

where, as before, B_{ij} is the submatrix of A that we get when we delete the ith row and j^{th} column of A. Then we define an $n \times n$ matrix $C = (c_{ij})$ by

$$C = {}^t(b_{ij}) ,$$

that is, $c_{ij} = b_{ji}$. This matrix C is called the *classical adjoint* of A, and we will denote it by adj A.

Example

$$\begin{aligned} n &= 2 \\ A &= \begin{pmatrix} a_{11} & a_{12} \\ a_{21} & a_{22} \end{pmatrix} \\ B_{11} &= a_{22}, \quad B_{12} = a_{21}, \quad B_{21} = a_{12}, \quad B_{22} = a_{11}. \end{aligned}$$

Then adj $A = \begin{pmatrix} a_{22} & -a_{12} \\ -a_{21} & a_{11} \end{pmatrix}$. Consider the product

$$A(\text{adj } A) = \begin{pmatrix} a_{11}a_{22} - a_{12}a_{21} & 0 \\ 0 & -a_{21}a_{12} + a_{11}a_{22} \end{pmatrix}.$$

So $A(\text{adj } A) = (\det A)I$. Similarly, $(\text{adj } A)A = (\det A)I$. So, at least in the case $n = 2$, when $\det A \neq 0$ we get A^{-1} easily from adj A. This is true in general.

Theorem 2 $A(\textit{adj } A) = (\textit{adj } A)A = (\det A)I$.

Corollary *If* $\det A \neq 0$, *then* $A^{-1} = (\frac{1}{\det A})\textit{adj } A$.

Proof: Now Theorem 2 makes an assertion about the matrix product $AC = A(\text{adj } A)$, and formula (22) says that the 1,1 entry in this product is $\det A$. To prove the theorem, we must show that the remaining diagonal entries in AC are $\det A$ and all nondiagonal entries are zero.

First we calculate the Laplace expansion about the ith row of A to get

$$\begin{aligned} \det A &= (-1)^{i+1}a_{i1} \det B_{i1} + \cdots + (-1)^{i+n}a_{in} \det B_{in} \\ &= a_{i1}b_{i1} + \cdots + a_{in}b_{in} \\ &= a_{i1}c_{1i} + \cdots + a_{in}c_{ni} \end{aligned} \tag{24}$$

which is just the i, i entry in AC.

Now suppose $i \neq k$, and we write an expression similar to the right hand side of (24):

$$(-1)^{k+1} a_{i1} \det B_{k1} + \cdots + (-1)^{k+n} a_{in} \det B_{kn}. \tag{25}$$

We recognize (25) as the i, k entry of AC. But it is also $\det D$ where D is the matrix obtained by replacing the kth ($\neq i^{th}$) row of A by the ith row (and leaving the ith row intact). Of course, since D has two identical rows, $\det D = 0$ by (ii) of Proposition 7.

The proof that $CA = I$ is almost exactly the same (starting off with (23) instead of (22)). So we have proven Theorem 2, and the corollary tells us A^{-1} whenever $\det A \neq 0$. ∎

Exercise. Calculate (25) for $n = 3, i = 1, k = 2$. Show it gives the $1, 2$ term of AC and it also gives

$$\det \begin{pmatrix} a_{11} & a_{12} & a_{13} \\ a_{11} & a_{12} & a_{13} \\ a_{31} & a_{32} & a_{33} \end{pmatrix}.$$

Systems of Equations

Let k be a field of characteristic zero. Let A be an $n \times m$ matrix over k and let $y \in k^m$. We are concerned with the questions:

(a) Does there exist $x \in k^n$ such that

$$xA = y? \tag{26}$$

(b) When such an x does exist, is it unique?

We can write (26) as

$$(x_1, x_2, \ldots, x_n) \begin{pmatrix} a_{11} & \cdots & a_{1m} \\ \vdots & & \\ a_{n1} & \cdots & a_{nm} \end{pmatrix} = (y_1, y_2, \ldots, y_m)$$

or even

$$(*) \begin{cases} a_{11}x_1 + \cdots + a_{n1}x_n = y_1 \\ \vdots \\ a_{1m}x_1 + \cdots + a_{nm}x_n = y_m. \end{cases}$$

The a_{ij} are the *coefficients* and $x_1, \ldots, x_n$ are the *unknowns*. We have m equations in n unknowns. Note that if $y = 0$, then we always have the *trivial solution* $x = 0$.

Now, using standard bases for k^n and k^m, we have

$$A : k^n \to k^m \qquad \text{linear.} \tag{26$'$}$$

The answer to (a) is: yes $\Leftrightarrow y \in A(k^n)$. (Notice that in (26), we have matrix multiplication so we write A on the right; but in (26$'$), we are considering A to be a linear map so we write A on the left, as in $A(k^n)$.) So if $n < m$, there will be some y's not in the image $A(k^n)$ and $xA = y$ will have no solution for such y's.

The answer to (b) is: yes $\Leftrightarrow \ker A = \{0\}$. That is, if $\ker A \neq \{0\}$, then no solution can be unique; for if $xA = y$ and $v \in \ker A$, then $(x+v)A = xA + 0 = y$. Conversely, if $x, x' \in k^n$ are distinct elements such that $xA = x'A = y$, then $(x - x')A = y - y = 0$ so that $x - x' \in \ker A$. In particular, if $n > m$, we must have $\ker A \neq \{0\}$.

So the interesting case is $n = m$. In this case, if $\det A \neq 0$, then A is an isomorphism and there is a unique solution. If $\det A = 0$, then the answer to both questions is no.

The previous discussion is interesting from a theoretical point of view, but it sheds little light on how to actually solve the system $(*)$, if a solution exists.

Recall that in section IA we briefly discussed the *method of elimination* when we solved a system of three equations in three unknowns. We now extend this method to the general system $(*)$ by using elementary row operations to simplify the system. (The elementary row operations E_1, E_2, E_3 are discussed at length in the exercises of the previous section.)

We begin by writing $(*)$ as an $m \times (n+1)$ matrix

$$\left(\begin{array}{ccc|c} a_{11} & \cdots & a_{n1} & y_1 \\ \vdots & & & \vdots \\ a_{1m} & \cdots & a_{nm} & y_m \end{array}\right)$$

and then putting this matrix into row reduced form using E_1, E_2, and E_3 (by exercise (13), we can always do this). From this new matrix, we can simply "read off" the solution, if it exists.

Some examples should suffice to illustrate this technique.

Examples

(1) Starting with the system

$$\begin{aligned} 3x_1 + 2x_2 &= 2 \\ 2x_1 + x_2 &= -1\ , \end{aligned}$$

we write this as

$$\left(\begin{array}{cc|c} 3 & 2 & 2 \\ 2 & 1 & -1 \end{array}\right)$$

Applying the elementary row operations to this matrix yields the row reduced matrix

$$\left(\begin{array}{cc|c} 1 & 0 & -4 \\ 0 & 1 & 7 \end{array}\right)$$

Thus the (unique) solution is $(x_1, x_2) = (-4, 7)$. Note that we expected this situation since

$$\det\begin{pmatrix} 3 & 2 \\ 2 & 1 \end{pmatrix} = 3 - 4 = -1 \neq 0 .$$

(2) Starting with the system

$$\begin{aligned} x_1 + x_2 &= 3 \\ 2x_1 + 3x_2 &= 4 \\ x_1 - 2x_2 &= -2 , \end{aligned}$$

we write this as

$$\left(\begin{array}{cc|c} 1 & 1 & 3 \\ 2 & 3 & 4 \\ 1 & -2 & -2 \end{array}\right)$$

which is row equivalent to

$$\left(\begin{array}{cc|c} 1 & 0 & 0 \\ 0 & 1 & 0 \\ 0 & 0 & 1 \end{array}\right) .$$

The third row of this matrix tells us that $0 = 0x_1 + 0x_2 = 1$; since this is clearly impossible, we conclude that no solution exists.

(3) Starting with the system

$$\begin{aligned} x_1 - x_2 + 2x_3 &= 1 \\ 2x_1 - x_2 + 3x_3 &= -1 , \end{aligned}$$

we obtain the row reduced matrix

$$\left(\begin{array}{ccc|c} 1 & 0 & 1 & -2 \\ 0 & 1 & -1 & -3 \end{array}\right) .$$

Thus we seek (x_1, x_2, x_3) such that

$$\begin{aligned} x_1 + x_3 &= -2 \\ x_2 - x_3 &= -3 , \end{aligned}$$

or equivalently,

$$\begin{aligned} x_1 &= -x_3 - 2 \\ x_2 &= x_3 - 3 . \end{aligned}$$

So any choice of $x_3 \in \mathbf{R}$ will determine x_1 and x_2 so that (x_1, x_2, x_3) is a solution.

It is an interesting fact that these examples exhaust all possibilities for the number of solutions to $(*)$; i.e., $(*)$ either has zero, one, or infinitely many solutions.

Exercises

(1) Find the inverses of the following matrices by using the corollary to Theorem 2.

(i) $A = \begin{pmatrix} 1 & 2 \\ -1 & 3 \end{pmatrix}$

(ii) $A = \begin{pmatrix} 1 & 0 & -1 \\ 2 & 1 & 0 \\ 0 & -1 & 1 \end{pmatrix}$

(2) If A is a nonsingular $n \times n$ matrix, show that A^{-1} can also be computed as follows. Form the $n \times 2n$ matrix

$$(A \mid I)$$

and row reduce this to

$$(I \mid B) .$$

Then $B = A^{-1}$.
(Hint: We know that A is the product of elementary matrices by a previous exercise.)

(3) Find A^{-1} where

$$A = \begin{pmatrix} 2 & 5 & -3 & -2 \\ -2 & -3 & 2 & -5 \\ 1 & 3 & -2 & 2 \\ -1 & -6 & 4 & 3 \end{pmatrix} .$$

(4) Solve the following systems of equations or indicate that no solution exists.

(i) $$\begin{aligned} 2x_1 + 3x_2 &= -1 \\ 5x_1 + 7x_2 &= 0 \end{aligned}$$

$$\text{(ii)} \quad \begin{aligned} x_1 - 2x_2 + 3x_3 &= 1 \\ 2x_1 + x_2 + 5x_3 &= -2 \end{aligned}$$

$$\text{(iii)} \quad \begin{aligned} x_1 - x_2 &= 4 \\ 2x_1 + x_2 &= 5 \\ -x_1 - 3x_2 &= -6 \end{aligned}$$

Homogeneous Systems.

Let A be an $n \times m$ matrix over k and consider the *homogeneous system* of linear equations

$$xA = 0$$

where $x \in k^n$. We know that $x = 0$ is a solution; what else can we say about the *solution space*

$$S = \{x \in k^n \mid xA = 0\} \ ?$$

(5) Show that S is a subsapce of k^n.

(6) Let W denote the row space of A, and show how we can consider $x \in S$ to be an element of the annihilator of W. Now show that $S \cong A(W)$.

(7) Show that $\dim S = \dim k^n - \dim W = n - \operatorname{rank}(A)$. (Hint: Recall exercise (7) of section I F.)

(8) What is the dimension of the solution space for the following homogeneous system?

$$\begin{aligned} 3x \ + 2y \ - z &= 0 \\ x \qquad\quad + z &= 0 \\ -2x \ - 3y \ + 2z &= 0 \end{aligned}$$

E. Eigenvalues

(Part ii)

Let V be a finite-dimensional vector space over a field k and let $\phi \in \operatorname{End}(V)$. Recall that $\lambda \in k$ is an eigenvalue for ϕ if there exists a nonzero vector $v \in V$ such that

$$\phi(v) = \lambda v.$$

Another way of saying this is to define $\psi_\lambda : V \to V$ by $\psi_\lambda(v) = \phi(v) - \lambda v$. Then ψ_λ is an endomorphism of V and clearly

(27) λ is an eigenvalue of $\phi \Leftrightarrow \ker \psi_\lambda \neq 0$.

Examples

(1) $V = \mathbf{R}^2$, $\phi = \begin{pmatrix} 0 & 1 \\ 0 & 0 \end{pmatrix}$.

For $\lambda \in \mathbf{R}$ and $v = a\,e_1 + b\,e_2$ consider

$$\psi_\lambda(v) = \phi(v) - \lambda v.$$

Since $\phi(e_1) = e_2$ and $\phi(e_2) = 0$, we get

$$\psi_\lambda(v) = a\,e_2 - \lambda(a\,e_1 + b\,e_2) = (-\lambda a)e_1 + (a - \lambda b)e_2.$$

If $\psi_\lambda(v)$ is to be 0, we must have $-\lambda a = 0$ and hence $a = 0$ or $\lambda = 0$. If $\lambda \neq 0$, then $a = 0$ and $\lambda b = 0$ so $b = 0$, and we conclude that $v = 0$. Thus ϕ has no nonzero eigenvalues. This is not surprising because ϕ is nilpotent (of order two) (i.e., $\phi \circ \phi = \phi^2 = \begin{pmatrix} 0 & 0 \\ 0 & 0 \end{pmatrix}$ the zero endomorphism). So if

$$\phi(v) = \lambda v, \quad 0 = \phi \circ \phi(v) = \phi(\lambda v) = \lambda\phi(v) = \lambda^2 v$$

and this cannot happen for $v \neq 0$ unless $\lambda = 0$. The same argument works for nilpotent endomorphisms of higher order.[1]

Exercise. Show that $\lambda = 0$ is, indeed, an eigenvalue for ϕ (in example (1)) and that e_2 is the only corresponding eigenvector (up to scalar multiplication).

(2) $V = \mathbf{R}^2$, $\phi = \begin{pmatrix} 0 & 0 \\ 0 & 1 \end{pmatrix}$.

Now $\phi(e_1) = 0$ and $\phi(e_2) = e_2$. So if $\lambda \neq 0$ and $v = a\,e_1 + b\,e_2$ and we evaluate $\psi_\lambda(v)$, we get $\psi_\lambda(v) = (-\lambda a)e_1 + (b - \lambda b)e_2$. If this is to be zero, we must have $a = 0$. If $v \neq 0$, then $b \neq 0$ and $(1 - \lambda)b = 0$ so we must have $\lambda = 1$. The eigenvector is any nonzero scalar times e_2. Here ϕ is idempotent ($\phi \circ \phi = \phi$) and does have exactly one nonzero eigenvalue.

Exercise. Show that $\lambda = 0$ is also an eigenvalue for ϕ (in example (2)) and find a corresponding eigenvector.

(3) $V = \mathbf{R}^2$, $\phi = \begin{pmatrix} 0 & 1 \\ -1 & 0 \end{pmatrix}$.

Proceeding as in (1) and (2), we get

$$\psi_\lambda(v) = (-b - \lambda a)e_1 + (a - \lambda b)e_2.$$

[1] e.g., if $\phi^3 = 0$ and $\phi(v) = \lambda v$, then $0 = \phi^3(v) = \lambda^3 v$ again implying $\lambda = 0$.

If this is to be zero, then $a = \lambda b$ and so

$$0 = b + \lambda a = b + \lambda^2 b = b(1 + \lambda^2).$$

Thus $b = 0$ and $a = 0$, and ϕ has no (real) eigenvalue. This is not surprising since ϕ is a rotation of $\frac{\pi}{2}$ radians.

Exercise. If we pass to the extension field $\mathbf{C}$ of $\mathbf{R}$, then we have $V = \mathbf{C}^2$ and $\lambda \in \mathbf{C}$. For ϕ (as in example (3)), find two distinct eigenvalues in this extension field. What are two corresponding eigenvectors?

(4) $V = \mathbf{R}^3$, $\lambda \in \mathbf{R}^* = \mathbf{R} - \{0\}$, $\phi = \begin{pmatrix} \lambda & a & b \\ 0 & \lambda & c \\ 0 & 0 & \lambda \end{pmatrix}$. We also assume $a, b, c \in \mathbf{R}^*$.

Now $\psi_\lambda = \phi - \lambda I = \begin{pmatrix} 0 & a & b \\ 0 & 0 & c \\ 0 & 0 & 0 \end{pmatrix}$, and we see that $e_3 \in \ker \psi_\lambda$. Thus by (27), λ is an eigenvalue for ϕ. Note that

$$\psi_\lambda^2 = \begin{pmatrix} 0 & 0 & ac \\ 0 & 0 & 0 \\ 0 & 0 & 0 \end{pmatrix},$$

and thus, although $\psi_\lambda(e_2) = (0, 0, c) \neq (0, 0, 0)$, we have $\psi_\lambda^2(e_2) = (0, 0, 0)$. So $\ker \psi_\lambda \subsetneq \ker \psi_\lambda^2$. We have

$$\begin{array}{ccccccc} 0 & \subsetneq & \ker \psi_\lambda & \subsetneq & \ker \psi_\lambda^2 & \subsetneq & \ker \psi_\lambda^3 = \mathbf{R}^3 . \\ & & \| & & \| & & \| \\ & & \mathrm{Span}(e_3) & & \mathrm{Span}(e_2, e_3) & & \mathrm{Span}(e_1, e_2, e_3) \end{array}$$

Exercise. Show that if we allow $\lambda = 0$ (in example (4)), then ϕ is nilpotent of order 3.

In example (2) above $\left(V = \mathbf{R}^2, \phi = \begin{pmatrix} 0 & 0 \\ 0 & 1 \end{pmatrix}\right)$ we have that the only nonzero eigenvalue is 1, and we find that if $v = ae_1 + be_2$, then $\psi_1(v) = -ae_1$ and thus $v \in \ker \psi_1 \Leftrightarrow a = 0$.

Now

$$\begin{aligned} \psi_1^2(v) &= \phi(\psi_1(v)) - \psi_1(v) \\ &= \phi(-a\, e_1) - (-a\, e_1) \\ &= 0 + a\, e_1 . \end{aligned}$$

So in example (2), $\ker \psi_1 = \ker \psi_1^2 (= \mathrm{Span}\ (e_2))$ and the dimensions of the kernels of ψ_1^m stabilize at 1. In example (4), the dimensions increased right up to the dimension of $V (= \mathbf{R}^3)$.

Definition The *algebraic multiplicity* of an eigenvalue λ of an endomorphism $\phi : V \to V$ is the maximum dimension of the subspaces $0 \subseteq \ker \psi_\lambda \subseteq \ker \psi_\lambda^2 \subseteq \cdots$. Since we are assuming V is finite-dimensional, this will be an integer $\leq \dim V$. If the maximum dimension is zero, this just means that λ is not an eigenvalue of ϕ.

Recall that back in Section E of Chapter I, for each $\lambda \in k$ we defined a subspace $V(\lambda)$ of V consisting of all $v \in V$ which had λ as an eigenvalue. We proved that if $\mu \neq \lambda$, then $V(\mu) \cap V(\lambda) = \{0\}$. In our present notation, $V(\lambda) = \ker \psi_\lambda$, and now we sometimes have bigger spaces (i.e., $\ker \psi_\lambda^i$) associated with λ. We will now show that for $\lambda \neq \mu$, these spaces also intersect only in zero.

To simplify our notation, we fix ϕ, λ, and μ (where $\lambda \neq \mu$). Then we write K_i for $\ker \psi_\lambda^i$ and L_j for $\ker \psi_\mu^j$ where $i, j = 1, 2, \ldots$. Note that we have $0 \subseteq K_1 \subseteq K_2 \subseteq \cdots \subseteq K_i \subseteq \cdots$ and $0 \subseteq L_1 \subseteq L_2 \subseteq \cdots \subseteq L_j \subseteq \cdots$.

We now claim that $K_i \cap L_j = \{0\}$ for each i, j.

Proof: Since $\psi_\lambda = \phi - \lambda I$ and ψ_λ^i is the zero endomorphism on K_i, we have that $(\phi - \lambda I)^i = 0$ on K_i. Similarly, $(\phi - \mu I)^j = 0$ on L_j.

Note that (since $\lambda \neq \mu$) the polynomials $(x - \lambda)^i$ and $(x - \mu)^j$ have no common factor. We will see in section B of the next chapter that the previous statement has the following consequence: we can find polynomials $a(x)$, $b(x)$ such that

$$1 = a(x)(x - \lambda)^i + b(x)(x - \mu)^j \,. \tag{28}$$

Now suppose we have $v \in K_i \cap L_j$. We wish to show that v must be 0. By (28), we have

$$\begin{aligned} v = I(v) &= [a(\phi)(\phi - \lambda I)^i + b(\phi)(\phi - \mu I)^j](v) \\ &= a(\phi)(\phi - \lambda I)^i(v) + b(\phi)(\phi - \mu I)^j(v) \\ &= 0 + 0 = 0 \,. \quad \blacksquare \end{aligned}$$

Note that everything we have said in this section about $\phi \in \mathrm{End}(V)$ applies equally to $A \in M_n(k)$; e.g., $\lambda \in k$ is an eigenvalue of the $n \times n$ matrix A if there is a nonzero $x \in k^n$ such that $xA = \lambda x$. We will soon see an easy way to calculate the eigenvalues of A.

Exercises

(1) Given any $\lambda \in k$, show that ϕ commutes with $\psi_\lambda = \phi - \lambda I$.

(2) Show that $K_i = \ker \psi_\lambda^i$ is ϕ-stable.

(3) Let $v_1, \ldots, v_m$ be nonzero eigenvectors for $\phi \in \mathrm{End}(V)$ corresponding to the distinct eigenvalues $\lambda_1, \ldots, \lambda_m$. Show that $v_1, \ldots, v_m$ are linearly independent.

(4) For $V = \mathbf{R}^3$ and A as below, find the eigenvalues of A along with the geometric and algebraic multiplicities of each.

$$\text{(i) } A = \begin{pmatrix} \lambda_1 & 1 & 0 \\ 0 & \lambda_1 & 0 \\ 0 & 0 & \lambda_2 \end{pmatrix}, \quad \lambda_1 \neq \lambda_2$$

$$\text{(ii) } A = \begin{pmatrix} 3 & 1 & 1 \\ 2 & 4 & 2 \\ 1 & 1 & 3 \end{pmatrix}$$

Chapter III

Rings and Polynomials

A. Rings

A *ring* R is a generalization of a field. It has operations of addition and multiplication, and R must be an abelian group under addition (just as for a field). However, the only requirement for multiplication is that it distribute over addition:

(1) $$a(b+c) = ab + ac, \text{ and } (a+b)c = ac + bc\,.$$

Example

If G is an abelian group (with identity element 0), we can make G into a ring by using the operation on G for addition and defining $ab = 0$ for all $a, b \in G$.

Definition A ring R is *associative* if its multiplication is associative; it is *commutative* if its multiplication is commutative; it has a *unit element* if it has a 2-sided multiplicative identity (i.e., a $1 \in R$ such that for any $x \in R$ we have $x1 = 1x = x$). If R has a unit element, then $a \in R$ is called a *unit* if there exists $b \in R$ such that $ab = 1 = ba$.

If R has a unit element, it is unique.

Examples

(1) The ring $\mathbf{Z}$ of integers is associative and commutative and has a unit element.

(2) The ring $M_2(\mathbf{R})$ of 2×2 real matrices is associative and has a unit element, but it is not commutative. The same is true of $M_n(k)$, for any $n > 1$ and any field k.

In Chapter V we will encounter some non-associative rings. But for this chapter we will assume that *ring* means *associative ring*.

Definition A subset S of a ring R is a *subring* of R if the operations on R induce operations on S so that S becomes a ring.

Example

In the ring $\mathbf{Z}$ of integers, the subset $2\mathbf{Z}$ of all even integers is a subring.

Definition A nonempty subset S of a ring $\mathbf{R}$ is an *ideal* if:

(i) $r \in R$ and $s \in S \Rightarrow rs \in S$ and $sr \in S$, and

(ii) $s, t \in S \Rightarrow s - t \in S$.

Let S be an ideal in R. Since S is nonempty, there is some $s \in S$ and thus $s - s = 0 \in S$. Thus every ideal contains zero. We also note that the single element $\{0\}$ is itself an ideal. ((ii) is obvious and (i) is just the proof in Chapter 0 that $a0 = 0$.)

Suppose R has a unit element 1. Then any ideal S in R containing 1 must equal R. (By (i), $r \in R \Rightarrow r1 (= r) \in S$.) For a field k, only $\{0\}$ and k are ideals. Because if I is an ideal in k, either $I = \{0\}$ or I contains a nonzero element x. In the latter case, x has an inverse $x^{-1} \in k$ and (by (i)) $x^{-1}x = 1 \in I$ implying $I = k$.

Definition If R, T are rings, a map

$$\phi : R \to T$$

is a (ring) *homomorphism* if for all $r, s \in R$ we have

(i) $\phi(r + s) = \phi(r) + \phi(s)$,

(ii) $\phi(rs) = \phi(r)\phi(s)$, and

(iii) if R, T have unit elements $1, 1'$ we also require that $\phi(1) = 1'$.

Note that a ring homomorphism sends the zero of R to the zero of T.

Proposition 1 *If $\phi : R \to T$ is a ring homomorphism, then*

(i) *$\phi(R)$ is a subring of T, and*

(ii) *$\ker\phi$ is an ideal in R.*

Proof: If $a, b \in \phi(R)$, then there exist elements x, y in R such that

$$\phi(x) = a \text{ and } \phi(y) = b.$$

Then

$$\phi(x \pm y) = a \pm b \text{ implying } a \pm b \in \phi(R), \text{ and}$$
$$\phi(xy) = ab \text{ implying } ab \in \phi(R).$$

This proves part (i).

For (ii), suppose $x \in \ker \phi$ and $r \in R$. Then $\phi(rx) = \phi(r)0 = 0$, so $rx \in \ker \phi$. Similarly, $xr \in \ker \phi$. For $x, y \in \ker \phi$, clearly $x - y \in \ker \phi$. ∎

Just as we did for vector spaces, we say that two rings R, T are *isomorphic* if there is a ring homomorphism $\phi : R \to T$ which is both surjective (i.e., $\phi(R) = T$) and injective (i.e., $\ker \phi = \{0\}$).

Note that in any ring R, any element $a \in R$ commutes with itself, so a^2 is well-defined. By associativity, $aa^2 = a^2a$ so a^3 is well-defined, etc., so all powers of a a are well-defined. In particular, for $A \in M_n(k)$, all powers A^p are well-defined. The ring $M_n(k)$ is also a vector space over the field k, and we have seen that the dimension of this vector space is n^2. Thus, of the vectors

$$\{I, A, A^2, A^3, \ldots\}$$

at most n^2 of them can be linearly independent. This means that there is a nontrivial linear combination of $I, A, A^2, \ldots, A^{n^2}$ which equals the zero matrix 0. That is, there exist $a_0, a_1, \ldots, a_{n^2} \in k$, which are not all zero, such that

$$a_0 I + a_1 A + a_2 A^2 + \cdots + a_{n^2} A^{n^2} = 0. \tag{2}$$

We rephrase this as follows: given $A \in M_n(k)$ there exists a nonzero polynomial (with coefficients in k) of degree (at most) n^2 which is satisfied by A. For $n = 2$ and $A = \begin{pmatrix} a_{11} & a_{12} \\ a_{21} & a_{22} \end{pmatrix}$ this means there is a nonzero vector

$$(b_0, b_1, b_2, b_3, b_4) \in k^4$$

such that

$$b_0 I + b_1 A + b_2 A^2 + b_3 A^3 + b_4 A^4 = \begin{pmatrix} 0 & 0 \\ 0 & 0 \end{pmatrix}.$$

Actually, it is possible to find a polynomial of degree n which $A \in M_n(k)$ satisfies.

Definition The *characteristic polynomial* of $A \in M_n(k)$ is

$$\sigma(x) = \det(xI - A) \tag{3}$$

which is a monic polynomial of degree n.

Strictly speaking, we have not defined the determinant of $C \in M_n(R)$ unless R is a field; however, we can formally calculate $\det C$, using the Laplace expansion, whenever C has entries from a commutative ring.

In particular, $\det C$ is defined for $C \in M_n(k[x])$. Taking $C = xI - A$, we see that (3) makes sense for $A \in M_n(k)$. (Actually, (3) makes sense for

$A \in M_n(R)$, where R is a commutative ring with unit element, but we will not need this much generality until Section C.)

Now the Cayley-Hamilton Theorem states that $\sigma(A)$ is the $n \times n$ zero matrix. In more detail,

$$xI - A = \begin{pmatrix} x - a_{11} & -a_{12} & \dots & -a_{1n} \\ -a_{21} & x - a_{22} & \dots & -a_{2n} \\ \vdots & & & \\ -a_{n1} & -a_{n2} & \dots & x - a_{nn} \end{pmatrix}$$

and the determinant $\sigma(x)$ of this matrix is of the form

$$\sigma(x) = x^n + b_{n-1}x^{n-1} + b_{n-2}x^{n-2} + \dots + b_1 x + b_0 , \tag{4}$$

and the assertion of the theorem becomes

$$A^n + b_{n-1}A^{n-1} + b_{n-2}A^{n-2} + \dots + b_1 A + b_0 I = 0.$$

For $n = 2$ and $A = \begin{pmatrix} a_{11} & a_{12} \\ a_{21} & a_{22} \end{pmatrix}$ the assertion is

$$A^2 - (a_{11} + a_{22})A + (a_{11}a_{22} - a_{12}a_{21})I = \begin{pmatrix} 0 & 0 \\ 0 & 0 \end{pmatrix} ,$$

which can be checked directly.

Note that $\sigma(0) = \det(-A) = (-1)^n \det A$, but $\sigma(0) = b_0$ is also the constant term of $\sigma(x)$. It is also an interesting fact that $b_{n-1} = -\text{trace}A$. Exercise. Use induction to show that $b_{n-1} = -\text{trace } A$ and that $\sigma(x)$ is, indeed, a monic polynomial of degree n.

Proposition 2 *$x - r$ is a factor of $\sigma(x) \Leftrightarrow r$ is an eigenvalue of A.*

Proof: $x - r$ is a factor of $\sigma(x) \Leftrightarrow \sigma(r) = 0 \Leftrightarrow \det(rI - A) = 0 \Leftrightarrow rI - A$ is singular $\Leftrightarrow$ there is a $v \neq 0$ in k^n such that $v(rI - A) = 0 \Leftrightarrow vA = rv$. ∎

Proposition 2 provides us with a convenient method for calculating the eigenvalues of a square matrix.

Examples

(1) The eigenvalues of $A = \begin{pmatrix} 2 & 2 \\ 2 & -1 \end{pmatrix}$ are the roots of

$$\sigma(x) = \det \begin{pmatrix} x - 2 & -2 \\ -2 & x + 1 \end{pmatrix} = (x-2)(x+1)-4 = (x-3)(x+2)$$

which are $\lambda_1 = 3$ and $\lambda_2 = -2$. To get an eigenvector for A corresponding to $\lambda_1 = 3$, note that we are looking for $(x_1, x_2) \in \mathbf{R}^2$ such that

$$(x_1, x_2)A = 3(x_1, x_2)$$

or equivalently

$$(x_1, x_2)(3I - A) = 0 \, .$$

Thus (x_1, x_2) must satisfy

$$(x_1, x_2)\begin{pmatrix} 1 & -2 \\ -2 & 4 \end{pmatrix} = (x_1 - 2x_2, -2x_1 + 4x_2) = (0, 0)$$

or, more simply, $x_1 = 2x_2$. If we choose $x_2 = 1$, then we get the eigenvector

$$(x_1, x_2) = (2, 1) \, .$$

Exercise. Find an eigenvector corresponding to the eigenvalue $\lambda_2 = -2$.

(2) The eigenvalues of $A = \begin{pmatrix} 0 & 1 \\ -1 & 0 \end{pmatrix}$ are the roots of

$$\sigma(x) = \det\begin{pmatrix} x & -1 \\ 1 & x \end{pmatrix} = x^2 + 1 \, .$$

This polynomial does not have any real roots, but if we extend to the complex numbers (i.e., think of A as an element of $M_2(\mathbf{C})$), then we see that the eigenvalues of A are $\lambda_1 = i$ and $\lambda_2 = -i$. Unfortunately, the eigenvectors for A must also be complex.

Since any real polynomial can be factored completely over $\mathbf{C}$, the technique used in example (2) shows that every $A \in M_n(\mathbf{R})$ has an eigenvalue (although it may be a complex number).

It is easy to show that similar matrices have equal characteristic polynomials; so if V is a finite-dimensional vector space, then we define the characteristic polynomial of $\phi \in \mathrm{End}(V)$ to be the characteristic polynomial of $M_\alpha(\phi)$, where α is any basis for V. This allows us to calculate the eigenvalues of ϕ by calculating them for any $M_\alpha(\phi)$.

Exercises

(1) Recall the definition of $\mathbf{Z}_m$ (with the operations $+$ and $\cdot$) from the exercises of Chapter 0. Show that $\mathbf{Z}_m$ is a ring. Show that this ring is a field $\Leftrightarrow m$ is a prime number.

(2) Let R be a ring and choose any $r, s \in R$. Show that $r(-s) = -(rs) = (-r)s$ and $(-r)(-s) = rs$. If R has a unit element, show that it is unique and that $(-1)r = -r$, $(-1)(-1) = 1$.

(3) A commutative ring is called an *integral domain* if it has no divisors of zero. Show that a finite integral domain is a field.

(4) Define a map $\phi : \mathbf{Z} \to \mathbf{Z}_m$ by $\phi(n)$ = remainder after dividing n by m. Show that ϕ is a surjective (ring) homomorphism. What is the kernel of ϕ?

(5) Find the eigenvalues of the following matrices. Also find a basis for each associated eigenspace.

$$\text{(i) } A = \begin{pmatrix} 1 & 0 \\ -1 & 2 \end{pmatrix}$$

$$\text{(ii) } A = \begin{pmatrix} 1 & 0 & 1 \\ 0 & 2 & 0 \\ 0 & 0 & 1 \end{pmatrix}$$

$$\text{(iii) } A = \begin{pmatrix} 2 & 2 & -1 \\ 1 & 1 & 1 \\ 1 & -2 & 4 \end{pmatrix}$$

(6) Show that similar matrices have the same characteristic polynomial.

(7) Show that the $n \times n$ matrices A and ${}^tA$ have the same characteristic polynomial. If one of the square matrices A, B is nonsingular, show that AB and BA have the same characteristic polynomial. Is this still true if we allow them to both be singular?

(8) Let $V = \mathbf{R}^3$ and define $\phi \in \operatorname{End}(V)$ by

$$\begin{aligned} \phi(e_1) &= 2e_1 + 2e_2 - e_3 \\ \phi(e_2) &= e_1 + e_2 + e_3 \\ \phi(e_3) &= e_1 - 2e_2 + e_4 . \end{aligned}$$

Find the eigenvalues of ϕ and find a basis for each associated eigenspace.

B. Polynomials

First we will consider polynomials with coefficients in a commutative (and associative) ring R with unit element. Then we will remove the commutativity assumption (since we will want to use $R = M_n(k)$) and see how this modifies our results.

Definition A function

$$p : R \to R$$

is a *polynomial function* if it is given by a formula of the sort

$$p(x) = a_n x^n + a_{n-1} x^{n-1} + \cdots + a_1 x + a_0 \tag{5}$$

where $a_n, \ldots, a_0$ are elements of R which we call the *coefficients* of p.

By writing (5) as we have, we mean to imply that $a_n \neq 0$. The n for this highest nonzero coefficient is the *degree* of the polynomial p and we write this as $\deg p$. The polynomial p is *monic* if $a_n = 1$. The x is referred to as the *indeterminate* (since it may be chosen to be any element of R) and the $a_i x^i$ are the *terms* of p.

Definition An element $x_0 \in R$ is a *root* of p if $p(x_0) = a_n x_0^n + \cdots + a_1 x_0 + a_0 = 0$.

Let $R[x]$ denote the set of all polynomials in x with coefficients in R. We will make $R[x]$ into a commutative ring with unit element.

The unit element 1 in $R[x]$ will be the polynomial in x with all coefficients except a_0 being zero and $a_0 = 1 \in R$. The zero $0 \in R[x]$ is the polynomial with all coefficients zero. (Note that we do not define the degree of the zero polynomial.)

To add $p(x)$ and $q(x)$ in $R[x]$ we simply add up all of their terms and then put the results in the form (5). For example, if $p(x) = 2x^3 + x - 1$ and $q(x) = 7x^4 - 3x^3 + x + 1$, we get $p(x) + q(x) = 7x^4 - x^3 + 2x$. This operation is associative and commutative, the polynomial 0 acts as identity element, and $-p(x)$ is the additive inverse of $p(x)$. Thus $R[x]$ becomes an abelian group under addition.

Since multiplication is to distribute over addition, we just need to define the product of a term $a_i x^i$ in $p(x)$ and a term $b_j x^j$ in $q(x)$. We set

$$(a_i x^i)(b_j x^j) = (a_i b_j) x^{i+j} \tag{6}$$

and extend by distributivity. This multiplication is associative and is commutative (since R is). Clearly $1 \in R[x]$ is the unit element.

Properties of Degree:

(a) $\deg(f+g) \leq \max(\deg f, \deg g)$

(b) $\deg(fg) \leq \deg f + \deg g$

If the coefficient of the highest power is a unit, for f or for g, then

(c) $\deg(fg) = \deg f + \deg g$

Proof: (a) and (b) are straightforward. For (c), let $f(x) = a_0 + a_1 x + \cdots + a_n x^n$ ($a_n \neq 0$) and $g(x) = b_0 + b_1 x + \cdots + b_m x^m$ ($b_m \neq 0$). Then we just need to show that the coefficient $a_n b_m$ of x^{n+m} is nonzero. But one of a_n, b_m is a unit. Say a_n is a unit. Then $a_n b_m = 0$ gives $a_n^{-1} a_n b_m = 0$ or $b_m = 0$, a contradiction. ∎

Proposition 3 (Division Theorem) *Given $f(x), g(x) \in R[x]$, with the highest coefficient in $g(x)$ being a unit, then there exist unique polynomials $q(x)$ and $r(x)$ such that*

$$f(x) = q(x)g(x) + r(x) \tag{7}$$

with $\deg r < \deg g$ *or* $r(x) = 0$.

Proof: (Note that if $\deg f < \deg g$, the conclusion is trivial — take $q(x) = 0$ and $r(x) = f(x)$.)

Let $S = \{f(x) - h(x)g(x) | h(x) \in R[x]\}$. If $0 \in S$, then we are done; otherwise, from S we select $r(x) = f(x) - q(x)g(x)$ with the least possible degree and suppose

$$\begin{aligned} r(x) &= c_0 + c_1 x + \cdots + c_p x^p, \qquad c_p \neq 0, \text{ and} \\ g(x) &= b_0 + b_1 x + \cdots + b_m x^m, \qquad b_m \text{ a unit.} \end{aligned}$$

We will show that $p \geq m$ contradicts the minimality of $p = \deg r$.

So suppose $p \geq m$, and let

$$s(x) = r(x) - c_p b_m^{-1} x^{p-m} g(x).$$

Then

$$\begin{aligned} s(x) &= f(x) - q(x)g(x) - c_p b_m^{-1} x^{p-m} g(x) \\ &= f(x) - (q(x) + c_p b_m^{-1} x^{p-m}) g(x) \end{aligned}$$

showing $s(x) \in S$. Now $\deg s < \deg r$ since

$$s(x) = (c_0 + c_1 x + \cdots + c_p x^p) - c_p b_m^{-1} x^{p-m} (b_0 + \cdots + b_m x^m)$$

and the pth powers cancel. Thus $\deg r < \deg g$, and we have also found q.

It remains to prove that q and r are unique. So suppose $qg + r = q'g + r'$ where $q \neq q'$. (Note that $q \neq q'$ implies $q - q' \neq 0$ so that $\deg(q - q') \geq 0$.) Then $(q - q')g = r' - r$. So

$$\deg((q - q')g) = \deg(r' - r) < \deg g. \tag{8}$$

By property (c) of degree,

$$\deg((q - q')g) = \deg(q - q') + \deg g, \tag{9}$$

and (8) and (9) together force $\deg(q - q') < 0$, giving a contradiction. Thus $q = q'$, and also $r = r'$. ∎

Proposition 4 *Let*

$$\begin{aligned} f(x) &= a_0 + a_1 x + \cdots + a_n x^n, \\ g(x) &= b_0 + b_1 x + \cdots + b_m x^m, \quad \textit{and } c \in R. \end{aligned}$$

Then

(i) $(fg)(c) = a_0g(c) + a_1g(c)c + \cdots + a_ng(c)c^n$, *and*

(ii) $(gf)(c) = g(c)a_0 + cg(c)a_1 + \cdots + c^ng(c)a_n$.

Proof: Using distributivity we get

$$f(x)g(x) = a_0g(x) + a_1g(x)x + \cdots + a_ng(x)x^n, \text{ and}$$
$$g(x)f(x) = g(x)a_0 + xg(x)a_1 + \cdots + x^ng(x)a_n,$$

so the proposition follows. (We have written these in a somewhat strange way in preparation for the case in which R is not necessarily commutative.) ∎

Corollary 1 *If* $g(c) = 0$, *then* $(fg)(c) = 0$ *and* $(gf)(c) = 0$.

Corollary 2 *If* $f(x)$ *can be written as* $f(x) = g(x)(x-c)+r$ *with* $c, r \in R$, *then* $f(c) = r$.

Corollary 3 (Factor Theorem) *Let* $f(x) \in R[x]$. *Then* $x - c$ *is a factor of* $f(x) \Leftrightarrow f(c) = 0$.

<u>Case of Noncommutative R</u>

Suppose now that R is not necessarily commutative, and let $p(x) = a_0 + a_1x + \cdots + a_nx^n \in R[x]$. We define operations on $R[x]$ as before and $R[x]$ becomes a ring which, of course, may not be commutative. Now for $c \in R$ we define the *value* of p at c to be

$$p(c) = a_0 + a_1c + \cdots + a_nc^n, \tag{10}$$

but in addition we define a *left value* $p_\lambda(c)$ by

$$p_\lambda(c) = a_0 + ca_1 + c^2a_2 + \cdots + c^na_n. \tag{11}$$

(Alternatively, we could write $p_\lambda(x) = a_0 + xa_1 + \cdots + x^na_n$ and evaluate $p_\lambda(x)$ at c.) Now Proposition 4 becomes

Proposition 4′

Let $f(x) = a_0 + a_1x + \cdots + a_nx^n$ *and* $g(x) \in R[x]$. *Then for* $c \in R$, *we have*

(i) $(fg)(c) = a_0g(c) + a_1g(c)c + \cdots + a_ng(c)c^n$, *and*

(ii) $(gf)_\lambda(c) = g_\lambda(c)a_0 + cg_\lambda(c)a_1 + \cdots + c^ng_\lambda(c)a_n$.

Proof: Again, we have

$f(x)g(x) = a_0 g(x) + a_1 g(x)x + \cdots + a_n g(x)x^n$ (recall our definition of the product of terms of polynomials. Thus $g(x)x = xg(x)$, etc.)

and

$$g(x)f(x) = g(x)a_0 + xg(x)a_1 + \cdots + x^n g(x)a_n.$$

Thus the proposition follows from (10) and (11) ∎

Corollary 1′

(*) $g(c) = 0 \Rightarrow (fg)(c) = 0$ (by (i) of 4′).

(**) $g_\lambda(c) = 0 \Rightarrow (gf)_\lambda(c) = 0$ (by (ii) of 4′).

Corollary 2′ (Remainder Theorem)

(i) If $f(x) = g(x)(x - c) + r$, then $f(c) = r$.

(ii) If $f(x) = (x - c)g(x) + r$, then $f_\lambda(c) = r$.

Proof: For (i), apply (*) to the product $g(x)(x - c)$. For (ii), apply (**) to the product $(x - c)g(x)$. ∎

Corollary 3′ (Factor Theorem)
Let $f(x) \in R[x]$ and $c \in R$. Then

(12) $(x - c)$ is a right factor of $f(x) \Leftrightarrow f(c) = 0$.

(13) $(x - c)$ is a left factor of $f(x) \Leftrightarrow f_\lambda(c) = 0$.

Polynomials Over a Field k

Definition If $p(x), q(x) \in k[x]$ we say that q *divides* p if there is some nonconstant $h(x) \in k[x]$ such that $p = qh$.

Given two polynomials $f(x), g(x)$ in $k[x]$ we now find their *greatest common divisor.*

Proposition 5 *Given $f(x), g(x) \in k[x]$ (not both zero), there is a unique polynomial $d(x)$ satisfying:*

(i) $d(x) = a(x)f(x) + b(x)g(x)$ *(for some $a(x), b(x) \in k[x]$),*

(ii) *$d(x)$ is monic,*

(iii) *$d(x)$ divides both $f(x)$ and $g(x)$, and*

(iv) *if $s(x)$ divides both $f(x)$ and $g(x)$, then $s(x)$ divides $d(x)$.*

Proof: Since we do not have both f and g zero, we have a nonzero polynomial of the form (i), and we may require it to be monic since k is a field. Let $d(x)$ be one of minimal degree. We claim that it then also satisfies (iii) and (iv).

If d fails to divide f, we have

$$f(x) = q(x)d(x) + r(x) \text{ with } \deg r < \deg d, \text{ and } r \neq 0.$$

For some $c \in k$, cr is monic. We claim it is of the form (i) and it has degree less than $\deg d$, contradicting the minimality of $\deg d$. Now $r = f - qd$ and $d = af + bg$, so

$$cr = cf - cqd = cf - cq(af + bg) = (c - cqa)f - cqbg \,,$$

and so cr is of the form (i). This proves d satisfies (iii).

(iv) follows easily from (i).

Finally, if we have d' with these same properties, then d and d' are monic polynomials each dividing the other (by part (iv)) and thus they are equal. ∎

Definition $f(x), g(x)$ are *relatively prime* if their greatest common divisor is 1. A polynomial f is *irreducible* if the only monic nonconstant polynomial dividing f is a constant multiple of f.

Lemma 1 *Suppose f is irreducible and f divides the product gh. Then f divides g or f divides h.*

Proof: If f does not divide g, then $1 = a(x)f(x)+b(x)g(x)$ (by Proposition 5). But then $h = afh + bgh$, and since f divides gh we conclude that f divides h. ∎

As a corollary to this lemma, if an irreducible $f(x) \in k[x]$ divides a multiple product of elements of $k[x]$, then it divides some one of them.

Proposition 6 *Let $p(x)$ be a nonconstant monic polynomial in $k[x]$. Then $p(x)$ is a unique (up to order) product of irreducible polynomials.*

Proof: Suppose monic nonconstant polynomials which cannot be factored into irreducible polynomials do exist. Let f be such a one of minimal degree. If f is irreducible, then it is already factored; so we must have $f = gh$ with g and h monic and nonconstant. But $\deg f = \deg g + \deg h$ (by (c) of properties of degree) and thus g and h can be factored into irreducible polynomials. But then f can also be factored.

Now suppose that there are polynomials which admit nonunique factorizations. Then we have

$$p = a_1 \ldots a_r = b_1 \ldots b_s \tag{14}$$

with each a_i and b_j monic and irreducible. Choose the least r for which this happens. Since a_1 divides $b_1 \dots b_s$, it divides one of the b_j; since b_j is irreducible, they are equal and can be cancelled and r was not minimal. Thus $r = s$.

Finally, if the b_j's are not just a permutation of the a_i's, we can choose $r(= s)$ minimal for this. Again, we can cancel a_1 with some b_j to obtain a contradiction. ∎

Definition Given $A \in M_n(k)$, we know that there is a monic polynomial $p(x) \in k[x]$ such that $p(A) = 0$ (e.g., let p be the characteristic polynomial of A) so there is such a polynomial of minimal degree. We call this the *minimal polynomial* of A and denote it by $m(x)$. (If V is finite-dimensional, we define the minimal polynomial of $\phi \in \text{End}(V)$ to be the minimal polynomial of any matrix which represents ϕ. This is well-defined by property (e) below.)

Properties of $m(x)$:

(a) $\deg m \leq n$

(b) m is unique

(c) m divides any polynomial which is satisfied by A (so, in particular, m divides the characteristic polynomial σ of A)

(d) m and σ have the same irreducible factors

(e) similar matrices have the same minimal polynomial

Proof: (a) is obvious. For (b), suppose we have another such $m'(x)$. Since m and m' are both monic, $m - m'$ must be of smaller degree; but $(m - m')(A) = m(A) - m'(A) = 0$ which contradicts the minimality of $\deg m = \deg m'$ (unless $m' = m$).

For (c), suppose we have a nontrivial polynomial $p(x) \in k[x]$ such that $p(A) = 0$. By the division theorem, there are polynomials $g(x), r(x)$ such that

$$p(x) = g(x)m(x) + r(x)$$

where $r = 0$ or $\deg r < \deg m$. If $r \neq 0$, we get a contradiction since $r(A) = p(A) - g(A)m(A) = 0$.

To prove (d), the following lemma will be useful. (Its proof is left as an exercise.)

Lemma 2 *$\sigma(x)$ divides $(m(x))^n$.*

Now suppose that $f(x) \in k[x]$ is an irreducible polynomial. If f divides m, then it also divides σ (since m divides σ by part (c)). On the other hand, suppose that f divides σ. Then f divides m^n by the previous lemma; but f is irreducible and so it must also divide m by Lemma 1.

The proof of (e) is left as exercise (2). ∎

Exercises

(1) Prove Lemma 2.

(2) Given $A \in M_n(k)$ and $f(x) \in k[x]$, show that $f(CAC^{-1}) = Cf(A)C^{-1}$ where C is any $n \times n$ invertible matrix. Conclude that $m(CAC^{-1}) = 0 \Leftrightarrow m(A) = 0$.

(3) Show that $x - 1$ divides $x^n - 1$.

(4) Given $A \in M_n(k)$, let $S \subseteq k[x]$ be the set of all polynomials $f(x)$ such that $f(A) = 0$. Show that S is an ideal of $k[x]$.

(5) Let $f(x) \in \mathbf{R}[x]$ be a monic irreducible polynomial of degree two. Show that f can be written in the form

$$f(x) = (x - a)^2 + b^2$$

where a, b are real numbers with $b \neq 0$. Conversely, show that any such f is irreducible.

(6) Suppose $f(x) = a_n x^n + \cdots + a_1 x + a_0$ is a polynomial with integer coefficients, and suppose m/ℓ is a root of f, where m and ℓ are relatively prime integers. Show that m divides a_0 and ℓ divides a_n.

(7) Show that the polynomials $x^2 - x + 4$ and $x^3 - x + 1$ are irreducible over the rational numbers. What about over $\mathbf{R}$?

(8) Show that if R is an integral domain with unit element, then any unit in $R[x]$ is also a unit in R.

(9) Find the minimal polynomial of the following matrices:

(i) $A = \begin{pmatrix} 2 & 1 \\ 0 & 2 \end{pmatrix}$

(ii) $A = \begin{pmatrix} 2 & 1 & 0 & 0 \\ 0 & 2 & 0 & 0 \\ 0 & 0 & 1 & 2 \\ 0 & 0 & 1 & 0 \end{pmatrix}$

(10) Suppose $A = \begin{pmatrix} B & 0 \\ 0 & C \end{pmatrix}$ where B, C are square submatrices. Show that the minimal polynomial of A is the least common multiple of the minimal polynomials for B and C. Now generalize to the case where

$$A = \begin{pmatrix} A_1 & & 0 \\ & \ddots & \\ 0 & & A_m \end{pmatrix}.$$

C. Cayley-Hamilton Theorem

Let R be a commutative ring with unit element. Then $R[x]$ is also commutative. We consider two rings. First consider $M_n(R[x])$, that is, the ring of all $n \times n$ matrices with entries from the ring $R[x]$. For example, $\begin{pmatrix} 1+x & 2x^2 \\ 3 & x^3 \end{pmatrix} \in M_2(\mathbf{Z}[x])$. Next consider $M_n(R)[x]$, that is, the ring of polynomials with coefficients from the ring $M_n(R)$ of $n \times n$ matrices with entries from R. We claim that these rings are essentially the same.

Theorem 1 $M_n(R[x]) \cong M_n(R)[x]$.

Proof: First we need some notation: for $f(x) = a_0 + a_1x + \cdots + a_nx^n \in R[x]$, let $r_p(f(x)) = a_p$, where it is understood that $r_p(f(x)) = 0$ if $p > n$.

Clearly $r_p(f(x) + g(x)) = r_p(f(x)) + r_p(g(x))$; since we want a ring isomorphism from $M_n(R[x])$ to $M_n(R)[x]$, we also need to know $r_p(f(x)g(x))$.

Lemma $r_p(f(x)(g(x)) = \sum_{\ell=0}^{p} r_\ell(f(x))r_{p-\ell}(g(x))$.

The proof of this lemma is left as an exercise.

Now for $F = (f_{ij}(x)) \in M_n(R[x])$, we define

$$r_p(F) = (r_p(f_{ij}(x)))$$

which is an element of $M_n(R)$. Clearly $r_p(F + G) = r_p(F) + r_p(G)$ for $F, G \in M_n(R[x])$.

We are now ready to define a map from $M_n(R[x])$ to $M_n(R)[x]$: given $F \in M_n(R[x])$, define

$$\alpha : M_n(R[x]) \to M_n(R)[x]$$

by $\alpha(F) = r_0(F) + r_1(F)x + r_2(F)x^2 + \ldots$ (a finite sum).

It is plain that $\alpha(F+G) = \alpha(F)+\alpha(G)$, since $r_p(F+G) = r_p(F)+r_p(G)$. To see that $\alpha(FG) = \alpha(F)\alpha(G)$ requires some work, however. We will show that the (matrix) coefficient of x^p is the same in $\alpha(FG)$ and $\alpha(F)\alpha(G)$.

The coefficient of x^p in $\alpha(F)\alpha(G)$ is

$$C = (c_{ij}) = \sum_{\ell=0}^{p} r_\ell(F)r_{p-\ell}(G)$$

by the obvious extension of our lemma to the polynomial ring $M_n(R)[x]$.

The coefficient of x^p in $\alpha(FG)$ is

$$D = (d_{ij}) = r_p(FG) .$$

If $F = (f_{ij}(x))$ and $G = (g_{ij}(x))$, then the i,j entry of FG is $(FG)_{ij} = f_{i1}g_{1j} + \cdots + f_{in}g_{nj}$. Thus

$$\begin{aligned} d_{ij} &= (r_p(FG))_{ij} = r_p(f_{i1}g_{1j} + \cdots + f_{in}g_{nj}) \\ &= \sum_{\ell=0}^{p}(r_\ell(f_{i1})r_{p-\ell}(g_{1j}) + \cdots + r_\ell(f_{in})r_{p-\ell}(g_{nj})) \end{aligned}$$

where the last equality follows from our lemma.

Now note that

$$\begin{aligned} c_{ij} &= \sum_{\ell=0}^{p}(i^{th} \text{ row of } r_\ell(F)) \times (j^{th} \text{ column of } r_{p-\ell}(G)) \\ &= \sum_{\ell=0}^{p}(r_\ell(f_{i1})r_{p-\ell}(g_{1j}) + \cdots + r_\ell(f_{in})r_{p-\ell}(g_{nj})) \end{aligned}$$

which is just d_{ij}.

We have shown that α is a ring homomorphism; the proofs that it is surjective and injective are left as an exercise. ∎

We continue with R (and hence $R[x]$) commutative and with unit element. Recall that we defined the characteristic polynomial of $A \in M_n(R)$ to be

(15) $$\sigma(x) = \det(xI - A).$$

We note that many important properties of the determinant function still hold in this general setting; in particular, Theorem 2 of Chapter II is still true. That is, if adj C denotes the classical adjoint of $C \in M_n(R)$, then

(16) $$(\text{adj } C)C = (\det C)I.$$

Applying this to $C = xI - A$ gives

$$(\text{adj } (xI - A))(xI - A) = \det(xI - A)I = \sigma(x)I.$$

Now we can consider (by Theorem 1) the relation

$$(\text{adj } (xI - A))(xI - A) = \sigma(x)I$$

to be in $M_n(R)[x]$. There, it says that $xI - A$ is a right factor of $\sigma(x)I$. Thus, by Corollary 3′, the value of $\sigma(x)I$ at $x = A$ is zero. We have proven the following important theorem.

Theorem 2 (Cayley-Hamilton Theorem) *For $A \in M_n(R)$, we have*

(17) $$\sigma(A) = a_0I + a_1A + \cdots + a_{n-1}A^{n-1} + A^n = 0$$

where $\sigma(x) = a_0 + a_1x + \cdots + a_{n-1}x^{n-1} + x^n$ is the characteristic polynomial of A.

Exercises

(1) Prove the lemma in this section. (Hint: Try induction on $\deg(fg)$.)

(2) Verify by direct computation that the following matrices are, indeed, roots of their characteristic polynomials.

$$\text{(i)} \quad A = \begin{pmatrix} 2 & 4 & -1 \\ -1 & 3 & 0 \\ 0 & 2 & 1 \end{pmatrix}$$

$$\text{(ii)} \quad A = \begin{pmatrix} 0 & 1 & 0 & 0 \\ 0 & 0 & 0 & 1 \\ 1 & 0 & 0 & 0 \\ 0 & 0 & 1 & 0 \end{pmatrix}$$

(3) Prove that α (defined in the proof of Theorem 1) is surjective and injective.

D. Spectral Theorems

(an application of the Cayley-Hamilton Theorem)

Let $A \in M_n(k)$ and let

$$\sigma(x) = \det(xI - A) \in k[x]$$

be its characteristic polynomial. We will see how to obtain information about A (considering it as an element of $\operatorname{End}(k^n)$) from information about σ.

Case I: Suppose

$$\sigma(x) = (x - r_1)(x - r_2)\dots(x - r_n)$$

where $r_i \in k$ and no two of the r_i equal.

First we note that by Cayley-Hamilton we have

$$0 = (A - r_1 I)(A - r_2 I)\cdots(A - r_n I)\,. \tag{18}$$

Next, since the r_i's are distinct, we can find $a_1, a_2, \dots, a_n \in k^*$ such that (for x not equal to any of the r_i's) we have

$$\frac{1}{\sigma(x)} = \frac{a_1}{x - r_1} + \cdots + \frac{a_n}{x - r_n}, \quad \text{or} \tag{19}$$

$$1 = a_1 \prod_{j \neq 1} (x - r_j) + \cdots + a_n \prod_{j \neq n} (x - r_j). \tag{20}$$

(This is just the technique of partial fractions from calculus.) Plugging in A for x, we have

$$I = a_1 \prod_{j \neq 1} (A - r_j I) + \cdots + a_n \prod_{j \neq n} (A - r_j I). \tag{21}$$

Setting $E_i = a_i \prod_{j \neq i} (A - r_j I)$, we have

$$I = E_1 + E_2 + \cdots + E_n. \tag{22}$$

Proposition 7 *(Properties of the E_i)*

(i) $E_i A = A E_i$ *(Each E_i commutes with A),*

(ii) $E_i A = r_i E_i$,

(iii) *For $i \neq j$, $E_i E_j = 0$ (The E_i's form an "orthogonal set"), and*

(iv) $E_i^2 = E_i$ *(Each E_i is* idempotent*).*

Proof:

(i) This follows since $A(A - r_j I) = (A - r_j I)A$ for each j.

(ii) $E_i(A - r_i I) = (a_i \prod_{j \neq i} (A - r_j I))(A - r_i I) = 0$ by (18).

(iii) Note that for $i \neq j$, the product

$$\left(a_i \prod_{\ell \neq i} (A - r_\ell I) \right) \left(a_j \prod_{m \neq j} (A - r_m I) \right)$$

contains the right hand side of (18) as a factor.

(iv)

$$\begin{aligned} E_i = E_i I &= E_i(E_1 + \cdots + E_n) \text{ by (22)} \\ &= E_i^2 \text{ by (iii).} \quad \blacksquare \end{aligned}$$

For $i = 1, \ldots, n$ we let

$$V_i = E_i(k^n) .$$

Since E_i is idempotent we know that it acts on V_i as the identity. Note that $E_i \neq 0$ since the minimal polynomial for A must contain the factor $x - r_i$. Hence $V_i \neq \{0\}$ and so $\dim V_i > 0$.

Proposition 8 (Spectral Theorem) *Each V_i is one-dimensional,*

$$k^n = V_1 \oplus \cdots \oplus V_n ,$$

and the action of A on V_i is multiplication by r_i. Furthermore, if we denote the restriction of A to V_i by A_i, then the minimal polynomial of A_i is $x - r_i$.

Proof: First off, the V_i do span k^n. For let $v \in k^n$. Then

$$v = vI = v(E_1 + \cdots + E_n) \in V_1 + V_2 + \cdots + V_n.$$

Next let $w = vE_i$ and consider the action of A on w: $wA = vE_iA = vr_iE_i = r_ivE_i = r_iw$. It follows that V_i is an eigenspace for A with eigenvalue r_i. Now suppose we have $w \in V_1 \cap (V_2 + \cdots + V_n)$ so that $w = v_1E_1$ and $w = v_2E_2 + \cdots + v_nE_n$. But then

$$w = wE_1 = (v_2E_2 + \cdots + v_nE_n)E_1 = 0$$

by (iii) of Proposition 7. Similarly, we see that $V_i \cap (V_1 + \cdots + V_{i-1} + V_{i+1} + \cdots + V_n) = \{0\}$ for each i. It follows that $k^n = V_1 \oplus \cdots \oplus V_n$ by exercise (4) of Section I D. Since $\dim V_i > 0$ for each i, we must have $\dim V_i = 1$.

The fact that $x - r_i$ is the minimal polynomial for A_i follows from (ii) of Proposition 7. ∎

So we see that Case I is very favorable. When $\sigma(x)$ splits into linear factors $x - r_i$ with no r_i occurring more than once, we get a clear and simple picture of A through the idempotents $E_1, \ldots, E_n$. In Case I it is also easy to calculate any polynomial $p(x)$ in A because we have

Proposition 9 $p(A) = p(r_1)E_1 + \cdots + p(r_n)E_n$.

Proof: By the division theorem and the remainder theorem, for each i we have $p(x) = p(r_i) + (x - r_i)q_i(x)$. By plugging A in for x and multiplying on the right by E_i, we have

$$p(A)E_i = p(r_i)E_i + (A - r_iI)q_i(A)E_i .$$

But $(A - r_iI)E_i = 0$ (by (ii) of Proposition 7). So

$$p(A)E_i = p(r_i)E_i.$$

Thus

$$\begin{aligned} p(A) &= p(A)(E_1 + \cdots + E_n) \\ &= p(r_1)(E_1) + \cdots + p(r_n)(E_n) \end{aligned}$$

as asserted. ∎

Case II: Suppose

$$\sigma(x) = \det(xI - A) = (x - r_1)^{m_1} \ldots (x - r_\ell)^{m_\ell} \tag{23}$$

with $r_1, \ldots, r_\ell$ distinct elements of the field k. So we are assuming again that $\sigma(x)$ splits into linear factors but now there may be repetitions. (For $k = \mathbf{C}$ we always have Case II and may sometimes have the subcase, Case I.)

By Cayley-Hamilton, we have

$$(24) \qquad 0 = (A - r_1 I)^{m_1} \ldots (A - r_\ell I)^{m_\ell}.$$

Just as for Case I, we want to find idempotents $E_1, \ldots, E_\ell$ associated with A by breaking up $1/\sigma(x)$ into an appropriate sum. The difference now is that we need polynomials as numerators instead of just elements of k.

Proposition 10 *If $p(x), q(x)$ are in $k[x]$ and are (nonconstant and) relatively prime, then there exist unique polynomials $a(x), b(x)$ in $k[x]$ such that $ap + bq = 1$ and $\deg a < \deg q$ and $\deg b < \deg p$.*

Proof: Since p, q are relatively prime, there exist $a_1, b_1 \in k[x]$ such that

$$a_1 p + b_1 q = 1.$$

Divide a_1 by q to get $a_1 = qh + a$ with $\deg a < \deg q$. Set $b = b_1 + hp$ and then $1 = a_1 p + b_1 q = (qh + a)p + (b - hp)q = ap + bq$. So for the existence part of the proof, we are done if we can show $\deg b < \deg p$.

Since we are in $k[x]$, with k a field, we always have $\deg(fg) = \deg f + \deg g$. So $\deg(bq) = \deg b + \deg q$. But we also have

$$\deg(bq) = \deg(1 - ap) \leq \deg a + \deg p < \deg q + \deg p$$

It follows that $\deg b < \deg p$, proving existence.

Suppose we also have $a'p + b'q = 1$ with $\deg a' < \deg q$ and $\deg b' < \deg p$. Then $(a - a')p = -(b - b')q$ and, since p and q are relatively prime, we have that p divides $b - b'$. Since $\deg(b - b') < \deg p$, this forces $b - b' = 0$. Similarly $a - a' = 0$ and Proposition 10 is proved. ∎

We are still considering Case II; i.e.,

$$\sigma(x) = \det(xI - A) = (x - r_1)^{m_1} \ldots (x - r_\ell)^{m_\ell}$$

with $r_1, \ldots, r_\ell$ distinct elements of k.

Proposition 11 *We have*

$$(25) \qquad \frac{1}{\sigma(x)} = \frac{a_1(x)}{(x - r_1)^{m_1}} + \cdots + \frac{a_\ell(x)}{(x - r_\ell)^{m_\ell}}$$

where $0 \leq \deg a_i(x) \leq m_i - 1$ (for $i = 1, \ldots, \ell$).

Proof: We will establish (25) in the equivalent form

$$1 = a_1(x)\prod_{j\neq 1}(x-r_j)^{m_j} + \cdots + a_\ell(x)\prod_{j\neq \ell}(x-r_j)^{m_j} \tag{26}$$

or equivalently, that

$$h(x) = a_1(x)\prod_{j\neq 1}(x-r_j)^{m_j} + \cdots + a_\ell(x)\prod_{j\neq \ell}(x-r_j)^{m_j} - 1$$

is the zero polynomial.

Now we choose $i \in \{1, 2, \ldots, \ell\}$ and set

$$p(x) = \prod_{j\neq i}(x-r_j)^{m_j} \text{ and } q(x) = (x-r_i)^{m_i}.$$

These are nonconstant relatively prime polynomials so we can apply Proposition 10. This gives unique polynomials $a_i(x), b_i(x)$ in $k[x]$ such that

$$a_i(x)p(x) + b_i(x)q(x) = 1 \tag{27}$$

with $\deg a_i < m_i$ and $\deg b_i < \deg p = \deg \sigma - m_i$. We use the $a_i(x)$ in (27) to define $h(x)$.

Note that $\deg h < \deg \sigma$. Also note that $q(x)$ divides $a_i(x)p(x) - 1$ (by (27)).

Now i was arbitrary so we conclude that for $(x - r_i)^{m_i}$ divides $h(x)$ for each i. But this implies $\sigma(x)$ divides $h(x)$. Since $\deg h < \deg \sigma$, we conclude that $h(x) = 0$ as we desired to prove. ∎

We can now substitute A for x in (26) to obtain

$$I = a_1(A)\prod_{j\neq 1}(A-r_jI)^{m_j} + \cdots + a_\ell(A)\prod_{j\neq \ell}(A-r_jI)^{m_j},$$

or letting $E_i = a_i(A)\prod_{j\neq i}(A-r_jI)^{m_j}$, we again have

$$I = E_1 + \cdots + E_\ell \tag{28}$$

with $E_iE_j = 0$ for $i \neq j$, $E_iA = AE_i$, and $E_i^2 = E_i$. So we have again decomposed I into a sum of "orthogonal" idempotents using A. However, condition (ii) in Proposition 7 (namely, that $(A - r_iI)E_i = 0$) must now be replaced by

$$(A - r_iI)^{m_i}E_i = 0. \tag{29}$$

So if we define $V_i = E_i(k^n)$ as before, we have that $V_i \neq \{0\}$ since the minimal polynomial of A must contain the factor $x - r_i$.

Proposition 12 (Spectral Theorem) *Each V_i is an m_i-dimensional subspace which is A-stable,*

$$k^n = V_1 \oplus \cdots \oplus V_\ell \,,$$

and the minimal polynomial of $A_i = A|_{V_i}$ is $(x - r_i)^{p_i}$ where $1 \leq p_i \leq m_i$. In fact, the minimal polynomial of A is $m(x) = (x - r_1)^{p_1} \cdots (x - r_\ell)^{p_\ell}$.

Proof: By (29), we conclude that the minimal polynomial of A_i must be $(x - r_i)^{p_i}$ with $a \leq p_i \leq m_i$. That $m(x) = (x - r_1)^{p_1} \cdots (x - r_\ell)^{p_\ell}$ is the minimal polynomial of A is left as an exercise. The proof that $k^n = V_1 \oplus \cdots \oplus V_\ell$ is almost identical to the proof in case I.

To see that V_i is A-stable, choose $w \in V_i$ and note that there is some $v \in k^n$ such that $w = vE_i$. Thus $wA = (vE_i)A = (vA)E_i \in V_i$.

We will defer the proof that $\dim V_i = m_i$ until the next section. (It is an easy consequence of the fact that we can put A into a "block diagonal form.") ∎

Exercises

(1) Compute the idempotents for the following matrices. What is the minimal polynomial of each $A_i = A|_{V_i}$?

(i) $$A = \begin{pmatrix} 1 & 0 & 0 \\ 0 & 2 & 0 \\ 0 & 0 & -1 \end{pmatrix}$$

(ii) $$A = \begin{pmatrix} 2 & 0 & 0 & 0 \\ 0 & 3 & 0 & 0 \\ 0 & 0 & -2 & 0 \\ 0 & 0 & 0 & 3 \end{pmatrix}$$

(2) Based on the results of exercise (1), state a general result about the idempotents of a diagonal matrix A.

(3) Let $A \in M_n(k)$ be the matrix with $\lambda \in k$ on its diagonal, 1 on its *superdiagonal*, and zeros elsewhere. For example, with $n = 3$

$$A = \begin{pmatrix} \lambda & 1 & 0 \\ 0 & \lambda & 1 \\ 0 & 0 & \lambda \end{pmatrix}.$$

What are the characteristic and minimal polynomials of A? Show that A has exactly one eigenvector. What is the algebraic multiplicity of the eigenvalue λ?

(4) Let V be a finite-dimensional vector space and suppose $V = V_1 \oplus V_2$ where V_1, V_2 are ϕ-stable subspaces for $\phi \in \mathrm{End}(V)$. If $m_i(x)$ is the minimal polynomial of $\phi|_{V_i}$, show that the minimal polynomial of ϕ is the least common multiple of $m_1(x)$ and $m_2(x)$. Generalize.

E. Jordan Form

Given $\phi \in \mathrm{End}(V)$ where V is finite-dimensional, it is to our advantage to choose a basis α of V so that $M_\alpha(\phi)$ is as simple as possible; this will make the task of understanding ϕ easier.

We will look at this primarily as a matrix problem; i.e., given $A \in M_n(k)$, we seek a matrix B which is similar to A and is as nice as possible.

Now suppose that the characteristic polynomial σ of A factors completely over k so that

$$\sigma(x) = (x - r_1)^{m_1} \cdots (x - r_\ell)^{m_\ell} \tag{30}$$

as in case II of the previous section. As we observed before, this is always true when $k = \mathbf{C}$.

By Proposition 12, there are A-stable subspaces $V_1, \ldots, V_\ell$ such that

$$k^n = V_1 \oplus \cdots \oplus V_\ell\,,$$

and the minimal polynomial of $A_i = A|_{V_i}$ is $(x - r_i)^{p_i}$ where $1 \le p_i \le m_i$. Also, the minimal polynomial of A is $m(x) = (x - r_1)^{p_1} \cdots (x - r_\ell)^{p_\ell}$.

First we show that we can consider A_i to be a square submatrix of A.

Proposition 13 *Suppose that ϕ is an endomorphism of the vector space V; also suppose that $W_1, \ldots, W_m$ are ϕ-stable subspaces such that $V = W_1 \oplus \cdots \oplus W_m$. Then we can find a basis α of V such that the matrix of ϕ is of the form*

$$M_\alpha(\phi) = \begin{pmatrix} B_1 & & & 0 \\ & B_2 & & \\ & & \ddots & \\ 0 & & & B_m \end{pmatrix} \tag{31}$$

where $B_1, \ldots, B_m$ are square submatrices. Furthermore, the matrix of $\phi|_{W_i}$ is B_i.

Proof: Choose a basis $\{v_1^1, \ldots, v_{n_1}^1\}$ for W_1. Since W_1 is ϕ-stable, we have

$$\begin{aligned} \phi(v_1^1) &= b_{11}v_1^1 + \cdots + b_{1n_1}v_{n_1}^1 \\ &\vdots \\ \phi(v_{n_1}^1) &= b_{n_1 1}v_1^1 + \cdots + b_{n_1 n_1}v_{n_1}^1\,. \end{aligned}$$

So let B_1 be the $n_1 \times n_1$ matrix (b_{ij}).

Similarly, choose a basis $\{v_1^2, \ldots, v_{n_2}^2\}$ for W_2, etc. Then $\alpha = \{v_1^1, \ldots, v_{n_m}^m\}$ is the desired basis for V. ■

Corollary *If $A \in M_n(k)$ satisfies (30), then there is a nonsingular matrix C such that $B = CAC^{-1}$ is of the form (31).*

Exercise. Finish the proof of Proposition 12; i.e., show that $\dim V_i = m_i$. (Hint: Show that the characteristic polynomial of A_i must be $(x - r_i)^{m_i}$.)

The matrix B in the preceding corollary is simpler than A since it is a *block diagonal matrix* (i.e., it has square submatrices on its diagonal and zeros elsewhere), but we can do better.

Definition The matrix

$$J_n(\lambda) = \begin{pmatrix} \lambda & 1 & & 0 \\ 0 & \lambda & \ddots & \\ \vdots & & \ddots & 1 \\ 0 & & & \lambda \end{pmatrix} \in M_n(k)$$

with λ on its diagonal, 1 on its superdiagonal, and zeros elsewhere is called the $n \times n$ *Jordan block* corresponding to λ.

Theorem 3 *Suppose that $\phi \in End(V)$ has $(x - r)^s$ as its minimal polynomial. Then there is a basis for V in which the matrix of ϕ is a block diagonal matrix of the form*

$$\begin{pmatrix} J_{s_1}(r) & & & 0 \\ & J_{s_2}(r) & & \\ & & \ddots & \\ 0 & & & J_{s_k}(r) \end{pmatrix} \tag{32}$$

where $s = s_1 \geq s_2 \geq \cdots \geq s_k \geq 1$. Furthermore, this matrix is uniquely determined by ϕ.

The proof of Theorem 3 is somewhat long-winded, so we will present it in an appendix after the exercises.

Returning to the case of a matrix $A \in M_n(k)$ whose characteristic polynomial is $\sigma(x) = (x - r_1)^{m_1} \cdots (x - r_\ell)^{m_\ell}$ and whose minimal polynomial is $m(x) = (x - r_1)^{p_1} \cdots (x - r_\ell)^{p_\ell}$, we see that a consequence of Proposition 13 and Theorem 3 is the following.

Corollary *There is a nonsingular matrix C such that $B = CAC^{-1}$ is the* *unique block diagonal matrix*

$$B = \begin{pmatrix} B_1 & & & 0 \\ & B_2 & & \\ & & \ddots & \\ 0 & & & B_\ell \end{pmatrix} \tag{33}$$

where B_i is the $m_i \times m_i$ block diagonal matrix of the form (32) corresponding to the eigenvalue r_i; furthermore, B_i has a $p_i \times p_i$ block (in its upper left hand corner) and none of its other blocks is larger.

Definition The matrix B in the preceding corollary is called the *Jordan form* of A.

Examples

(1) $A = \begin{pmatrix} 1 & 2 \\ 0 & 1 \end{pmatrix}$ has characteristic polynomial $\sigma(x) = (x-1)^2$ and minimal polynomial $m(x) = (x-1)^2$; so the Jordan form of A must have exactly one 2×2 Jordan block corresponding to the eigenvalue 1, i.e.,

$$B = \begin{pmatrix} 1 & 1 \\ 0 & 1 \end{pmatrix}.$$

(2) A is said to be *diagonalizable* if its Jordan form is a diagonal matrix. If we are in case I of the previous section, then A is diagonalizable.

(3) Suppose the minimal polynomial of $A \in M_4(\mathbf{R})$ is $(x-2)^2$. Then A has two possible Jordan forms, namely,

$$\begin{pmatrix} 2 & 1 & & \\ 0 & 2 & & \\ & & 2 & 1 \\ & & 0 & 2 \end{pmatrix} \quad \text{or} \quad \begin{pmatrix} 2 & 1 & & \\ 0 & 2 & & \\ & & 2 & \\ & & & 2 \end{pmatrix}.$$

(Here, a blank space indicates a zero.) In order to decide which is actually the Jordan form of A, we need more information about A.

We will now use the Jordan form of A to draw some interesting (and important) conclusions.

Since r_i is a root of the characteristic polynomial $\sigma(x)$ of A, it is also an eigenvalue of A. Recall that we defined the algebraic multiplicity of r_i to be the maximum dimension of the subspaces $0 \subseteq \ker\,(A - r_iI) \subseteq \ker\,(A - r_iI)^2 \subseteq \cdots$.

Proposition 14 *The algebraic multiplicity of r_i is equal to m_i (the multiplicity of r_i as a root of $\sigma(x)$).*

Proof: Let $B = CAC^{-1}$ be the Jordan form of A (as in (33)) so that B_i is the $m_i \times m_i$ block corresponding to r_i.

Now the only vectors which get mapped to 0 by $B - r_i I$ are precisely the ones which get mapped to 0 by $B_i - r_i I$. (Exercise. Show this.) Similarly, ker $(B - r_i I)^j =$ ker $(B_i - r_i I)^j$ for all $j \geq 1$. But $B_i - r_i I$ is nilpotent of order p_i (i.e., $(B_i - r_i I)^{p_i} = 0$) so the dimensions of the kernels of $(B - r_i I)^j$ stabilize at m_i.

Since dim(ker $(A - r_i I)^j$) = dim(ker $(B - r_i I)^j$) for all $j \geq 1$, our result follows. ∎

Recall that we defined the geometric multiplicity of the eigenvalue r_i to be the dimension of the eigenspace belonging to r_i.

Proposition 15 *If B_i is the block corresponding to r_i in the Jordan form of A, then the geometric multiplicity of r_i is equal to the number of Jordan blocks occurring on the diagonal of B_i.*

Proof: Exercise. (Hint: An equivalent formulation of this problem is to show that the dimension of ker $(B_i - r_i I)$ is equal to the number of Jordan blocks on the diagonal of B_i.) ∎

Note that if $B = (b_{ij})$ is a matrix in Jordan form, then its determinant is easy to calculate:

$$\det B = b_{11} b_{22} \cdots b_{nn}$$

since B is triangular.

For $A \in M_n(\mathbf{C})$, let $\lambda_1, \ldots, \lambda_n \in \mathbf{C}$ be its eigenvalues (there may be repeats in this list). If $B = CAC^{-1}$ is the Jordan form of A, then the diagonal elements of B are precisely (some permutation of) $\lambda_1, \ldots, \lambda_n$. Since $\det A = \det B$, we see that the determinant of A is simply the product of its eigenvalues.

In fact, for $A \in M_n(\mathbf{R})$, the same is true: thinking of A as an element of $M_n(\mathbf{C})$, we can put it in Jordan form (over $\mathbf{C}$), etc.

Similar arguments show that trace $A = \lambda_1 + \cdots + \lambda_n$ for A with real or complex entries.

We summarize these paragraphs as

Proposition 16 *Let $\lambda_1, \ldots, \lambda_n \in \mathbf{C}$ be the eigenvalues of $A \in M_n(k)$ where $k = \mathbf{R}$ or $k = \mathbf{C}$. Then $\det A = \lambda_1 \cdots \lambda_n$ and trace $A = \lambda_1 + \cdots + \lambda_n$.*

<u>Exercises</u>

(1) Find all possible Jordan forms for a real matrix whose characteristic and minimal polynomials are as follows. (Could Proposition 15 possibly be used to single out a particular Jordan form?)

(i) $\sigma(x) = (x-1)^4(x-2)^2$, $m(x) = (x-1)^2(x-2)^2$

(ii) $\sigma(x) = (x+3)^4(x-4)^4$, $m(x) = (x+3)^2(x-4)^2$

(iii) $\sigma(x) = (x+1)^6$, $m(x) = (x+1)^2$

(2) Find the Jordan form of the following matrices. Also calculate $\det A$ and trace A.

$$\text{(i) } A = \begin{pmatrix} 2 & 1 \\ -1 & 4 \end{pmatrix}$$

$$\text{(ii) } A = \begin{pmatrix} 1 & 1 & -1 \\ 2 & 2 & 1 \\ 2 & -1 & 4 \end{pmatrix}$$

$$\text{(iii) } A = \begin{pmatrix} 2 & 0 & & & \\ 5 & 2 & & & \\ & & 7 & & \\ & & & 4 & 2 \\ & & & 3 & 5 \end{pmatrix}$$

$$\text{(iv) } A = \begin{pmatrix} 1 & -1 & 0 \\ 2 & 1 & 3 \\ 1 & 2 & 0 \end{pmatrix}$$

$$\text{(v) } A = \begin{pmatrix} 1 & 2 & 3 & 4 \\ 0 & 1 & 2 & 3 \\ 0 & 0 & 1 & 2 \\ 0 & 0 & 0 & 1 \end{pmatrix}$$

(3) Show that $A \in M_n(k)$ is diagonalizable $\Leftrightarrow$ A has n linearly independent eigenvectors.

(4) Show that $A \in M_n(k)$ is diagonalizable $\Leftrightarrow$ the minimal polynomial $m(x)$ of A is a product of distinct linear factors (i.e., $m(x) = (x - r_1)\cdots(x - r_\ell)$ where $r_1, \ldots, r_\ell$ are distinct).

(5) Show that the Jordan block $J_n(r)$ is not diagonalizable. Which of the matrices in exercise (2) are diagonalizable?

(6) If $A \in M_n(k)$ is nilpotent and nonzero, show that A is not diagonalizable.

(7) If $A \in M_n(\mathbf{C})$ satisfies $A^m = I$, show that A is diagonalizable. (Hint: Show that $J_n(r)^m$ is not diagonal unless $n = 1$.)

(8) Given the characteristic and minimal polynomials of some $A \in M_3(\mathbf{C})$, show that these completely determine the Jordan form of A.

(9) Suppose that $A \in M_n(k)$ is diagonalizable, and let Q be the matrix whose rows are the n linearly independent eigenvectors of A. Show that $B = QAQ^{-1}$ is the Jordan form of A.

Appendix: Proof of Theorem 3

First we consider the special case where $\dim V = s$ and the minimal polynomial of ϕ is $(x-r)^s$. Since $(\phi - r)^{s-1} \neq 0$, there is a vector $v \in V$ such that $(\phi - r)^{s-1}(v) \neq 0$.

Observe that the set of s vectors

$$\alpha = \{v, (\phi - r)(v), \ldots, (\phi - r)^{s-1}(v)\}$$

is linearly independent. For if

$$a_0 v + a_1(\phi - r)(v) + \cdots + a_{s-1}(\phi - r)^{s-1}(v) = 0\,, \tag{34}$$

then applying $(\phi - r)^{s-1}$ to both sides of (34) yields

$$a_0(\phi - r)^{s-1}(v) = 0\,.$$

But then $a_0 = 0$ since $(\phi - r)^{s-1}(v) \neq 0$; and so (34) becomes

$$a_1(\phi - r)(v) + \cdots + a_{s-1}(\phi - r)^{s-1}(v) = 0\,. \tag{34'}$$

Applying $(\phi - r)^{s-2}$ to both sides of (34′) yields $a_1 = 0$, etc. Hence $a_0 = \cdots = a_{s-1} = 0$ and so α is linearly independent.

So α is a basis for V. What is the matrix of ϕ in this basis? Since

$$\begin{aligned}
&\phi(v) = (r + \phi - r)(v) = rv + (\phi - r)(v)\\
&\phi(\phi - r)(v) = (r + \phi - r)(\phi - r)(v) = r(\phi - r)(v) + (\phi - r)^2(v)\\
&\vdots\\
&\phi(\phi - r)^{s-1}(v) = r(\phi - r)^{s-1}(v) + r(\phi - r)^s(v) = r(\phi - r)^{s-1}(v)
\end{aligned}$$

we have $M_\alpha(\phi) = J_s(r)$.

Now we return to the general case where $\dim V = n$ and the minimal polynomial of ϕ is $(x-r)^s$. Suppose that we can decompose V as a direct sum of ϕ-stable subspaces

$$V = V_1 \oplus \cdots \oplus V_k \tag{35}$$

where $\dim V_i = s_i$ (remember that s_1 must be s) and the minimal polynomial of $\phi|_{V_i}$ is $(x-r)^{s_i}$. Then by the simple case above and Proposition 13, we will be done.

For convenience, we write

$$\psi = \phi - r \, .$$

We start off by finding V_1 as in the simple case. Since $\psi^{s-1} \neq 0$, there is a $v_1 \in V$ such that $\psi^{s-1}(v_1) \neq 0$. Let V_1 be the $s_1(= s)$-dimensional ϕ-stable subspace

$$V_1 = \mathrm{Span}(v_1, \psi(v_1), \ldots, \psi^{s-1}(v_1))$$

and note that the minimial polynomial of $\phi|_{V_1}$ is $(x-r)^{s_1}$.

If we can find a ϕ-stable subspace W such that

$$V = V_1 \oplus W \, ,$$

then (noting that the minimal polynomial of $\phi|_W$ is $(x-r)^t$ where $t \leq s$) by induction on $n = \dim V$, we can find the appropriate ϕ-stable subspaces $V_2, \ldots, V_k$ such that

$$W = V_2 \oplus \cdots \oplus V_k \, .$$

This will then give us (35).

To get W, we use induction on s. If $s = 1$, there is nothing to prove. Our induction hypothesis is:

> Suppose $\rho \in \mathrm{End}(\mathbf{X})$ has minimal polynomial $(x-r)^{s-1}$. If we let $\mathbf{X}_1 = \mathrm{Span}(x_1, (\rho - r)(x_1), \ldots, (\rho - r)^{s-2}(x_1))$ where $(\rho - r)^{s-2}(x_1) \neq 0$, then there is a ρ-stable subspace $\mathbf{Y}$ such that $\mathbf{X} = \mathbf{X}_1 \oplus \mathbf{Y}$.

Now let $V' = \psi(V)$ and note that V' is ϕ-stable, because for any $v \in V$ we have

$$\phi(\psi(v)) = \phi(\phi - r)(v) = (\phi - r)\phi(v) = \psi(\phi(v)) \in V' \, .$$

Also, the minimal polynomial of $\phi|_{V'}$ is $(x-r)^{s-1}$. If we let $u_1 = \psi(v_1)$, then

$$V_1' = \mathrm{Span}(u_1, \psi(u_1), \ldots, \psi^{s-2}(u_1)) = \psi(V_1)$$

is a ϕ-stable subspace of V', so we have a ϕ-stable subspace V_2' such that

$$V' = V_1' \oplus V_2'$$

by induction.

Now let

$$W' = \psi^{-1}(V_2') = \{v \in V \mid \psi(v) \in V_2'\}$$

and note that

$$V = V_1 + W' \, . \tag{36}$$

To see this, choose $v \in V$ and observe that we can write $\psi(v)$ as

$$\psi(v) = w_1 + w_2$$

where $w_i \in V_i'$. But $V_1' = \text{Span}(\psi(v_1), \ldots, \psi^{s-1}(v_1))$ so we can write w_1 as

$$\begin{aligned} w_1 &= a_1\psi(v_1) + \cdots + a_{s-1}\psi^{s-1}(v_1) \\ &= \psi(a_1v_1 + \cdots + a_{s-1}\psi^{s-2}(v_1)) \\ &= \psi(z_1) \end{aligned}$$

where $z_1 \in V_1$. Thus

$$\psi(v) = \psi(z_1) + w_2$$

so that $\psi(v - z_1) = w_2 \in V_2'$. But this implies that $v - z_1 \in W'$; so we have

$$v = z_1 + (v - z_1) \in V_1 + W'$$

as claimed.

Unfortunately, we do not have $V_1 \cap W' = \{0\}$, so the sum (36) is not direct. However, we can say that

$$V_1 \cap V_2' = \{0\}\ . \tag{37}$$

Exercise. Show this. (Hint: If $v \in V_1 \cap V_2'$, then $\psi(v) \in V_1' \cap V_2' = \{0\}$. Show that $\psi(v) = 0$ and $v \in V_1$ together imply that $v \in V_1'$.)

So W' contains the subspaces $V_1 \cap W'$ and V_2' which intersect trivially (by (37)). Thus we can find another subspace V_3' such that

$$W' = (V_1 \cap W') \oplus V_2' \oplus V_3'\ .$$

Exercise. Show this.

Finally, we let

$$W = V_2' \oplus V_3'$$

and show that this is the subspace we desire.

Clearly, $V_1 \cap W = \{0\}$ since V_2' and V_3' are both disjoint from V_1. Furthermore, $V_1 + W$ contains both V_1 and W' so that $V = V_1 + W$. Thus

$$V = V_1 \oplus W\ .$$

To see that W is ϕ-stable, choose $w \in W$ and note that $\psi(w) \in V_2'$. But then $\phi(w) = (\phi - r + r)(w) = \psi(w) + rw \in V_2' + W = W$.

We have shown that ϕ can be represented by the matrix (32), but what about uniqueness? Toward this end, the following lemma will be helpful.

Lemma *Suppose that U is t-dimensional and that $\rho \in End(U)$ has minimal polynomial x^t. Then $\rho^j(U)$ has dimension $t - j$ for each $j \leq t$.*

Proof: Choose $u \in U$ so that $\{u, \rho(u), \ldots, \rho^{t-1}(u)\}$ is a basis for U. Then $\{\rho^j(u), \ldots, \rho^{t-1}(u)\}$ is a basis for $\rho^j(U)$. ∎

Continuing the proof of uniqueness in Theorem 3, suppose that ϕ has another block diagonal representation where the blocks are $J_{t_1}(r), \ldots, J_{t_\ell}(r)$ with $t_1 \geq \cdots \geq t_\ell \geq 1$. (These blocks correspond to subspaces $W_1, \ldots, W_\ell$ such that $V = W_1 \oplus \cdots \oplus W_\ell$.) We claim that $k = \ell$ and $s_i = t_i$ for each i. If not, there is a least index j such that

$$s_j \neq t_j ;$$

without loss of generality, we assume that $s_j < t_j$.

As before, let $\psi = \phi - r$ and note that the minimal polynomial at $\psi|_{W_i}$ is x^{t_i} (and for $\psi|_{V_i}$ it is x^{s_i}). By the previous lemma,

$$\dim \psi^{s_j}(V_i) = s_i - s_j \quad \text{and} \quad \dim \psi^{s_j}(W_i) = t_i - s_j$$

for $i \leq j$.

By exercise (9) of Section I E, we have

$$\psi^{s_j}(V) = \psi^{s_j}(W_1 \oplus \cdots \oplus W_\ell) = \psi^{s_j}(W_1) \oplus \cdots \oplus \psi^{s_j}(W_\ell)$$

and so

$$(38) \qquad \begin{aligned} \dim \psi^{s_j}(V) &\geq \dim \psi^{s_j}(W_1) + \cdots + \dim \psi^{s_j}(W_j) \\ &= (t_1 - s_j) + \cdots + (t_j - s_j) \,. \end{aligned}$$

Also $\psi^{s_j}(V_i) = \{0\}$ for $i \geq j$, so we have

$$\psi^{s_j}(V) = \psi^{s_j}(V_1 \oplus \cdots \oplus V_k) = \psi^{s_j}(V_1) \oplus \cdots \oplus \psi^{s_j}(V_{s_{j-1}})$$

which implies that

$$(39) \qquad \dim \psi^{s_j}(V) = (s_1 - s_j) + \cdots + (s_{j-1} - s_j) \,.$$

Now $s_1 = t_1, \ldots, s_{j-1} = t_{j-1}$ by our choice of j, so we can rewrite (39) as

$$(39') \qquad \dim \psi^{s_j}(V) = (t_1 - s_j) + \cdots + (t_{j-1} - s_j) \,.$$

Comparing (38) and (39′), we conclude that

$$0 \geq t_j - s_j$$

which contradicts the fact that $s_j < t_j$. This completes the proof of Theorem 3. ∎

Chapter IV

Inner Product Spaces

A. $\mathbf{R}^n$ as a Model, Bilinear Forms

For $x = (x_1, \ldots, x_n)$, $y = (y_1, \ldots, y_n)$ two points in $\mathbf{R}^n$ we define the *distance* from x to y by

$$d(x,y) = \sqrt{(x_1 - y_1)^2 + \cdots + (x_n - y_n)^2}.$$

Two obvious properties of d are:

(1) $d(y,x) = d(x,y)$,

(2) $d(x,y) \geq 0$, and $d(x,y) = 0 \Leftrightarrow x = y$.

A less obvious property is:

(3) For any $x, y, z \in \mathbf{R}^n$ we have

$$d(x,y) + d(y,z) \geq d(x,z).$$

Property (3) is called the *triangle property.* A function $d : X \times X \to \mathbf{R}$ satisfying (1), (2), and (3) is a *metric* on the set X.

To prove (3), it is convenient to introduce some related definitions. The *norm* of $x \in \mathbf{R}^n$ is

$$\|x\| = d(x,0) = \sqrt{x_1^2 + \cdots + x_n^2}.$$

The function $\langle\, , \rangle : \mathbf{R}^n \times \mathbf{R}^n \to \mathbf{R}$ defined by

$$\langle x, y \rangle = x_1 y_1 + \cdots + x_n y_n \tag{1}$$

is called an *inner product.* We will see that d and $\langle\, , \rangle$ are related by

$$d(x,y) = \sqrt{\langle x - y, x - y \rangle}. \tag{2}$$

Proposition 1 $\langle\, ,\rangle$ *has the following properties:*

(i) $\langle\, ,\rangle$ *is* symmetric; *i.e.,* $\langle x,y\rangle = \langle y,x\rangle$,

(ii) $\langle x,y+z\rangle = \langle x,y\rangle + \langle x,z\rangle$,

(ii′) $\langle x+y,z\rangle = \langle x,z\rangle + \langle y,z\rangle$,

(iii) $\langle x,ay\rangle = a\langle x,y\rangle$ *for* $a \in \mathbf{R}$,

(iii′) $\langle ax,y\rangle = a\langle x,y\rangle$ *for* $a \in \mathbf{R}$,

(iv) $\langle\, ,\rangle$ *is* positive definite; *i.e.,* $\langle x,x\rangle$ *is always* ≥ 0, *and* $\langle x,x\rangle = 0 \Leftrightarrow x = 0 = (0,\ldots,0)$,

(v) *if* $\{e_1,\ldots,e_n\}$ *is the standard basis for* $\mathbf{R}^n$, *then*

$$\langle e_i,e_j\rangle = \delta_{ij} = \begin{cases} 1 & \text{if} \quad i=j \\ 0 & \text{if} \quad i \neq j\,, \text{and} \end{cases}$$

(vi) $\langle\, ,\rangle$ *is* nondegenerate; *i.e., if* $\langle x,y\rangle = 0$ *for all* y, *then* $x = 0$.

Proof: (i), (ii), (ii′), (iii), (iii′), and (iv) are all routine to prove. Note that (ii) and (iii) together say that $\langle\, ,\rangle$ is linear in the second variable when the first variable is held fixed. We have a similar statement from (ii′) and (iii′). These four together mean that $\langle\, ,\rangle$ is *bilinear.*

(v) is obvious. For (vi), suppose $\langle x,y\rangle = 0$ for all y. Then, letting $y = x$ and using (iv), we see that $x = 0$. ∎

Now we prove

$$d(x,y) = \sqrt{\langle x-y,x-y\rangle} = \|x-y\|. \tag{3}$$

Using bilinearity of $\langle\, ,\rangle$, we have

$$\begin{aligned} \langle x-y,x-y\rangle &= \langle x,x\rangle - \langle y,x\rangle - \langle x,y\rangle + \langle y,y\rangle \\ &= (x_1-y_1)^2 + \cdots + (x_n-y_n)^2 = \|x-y\|^2 \\ &= (d(x,y))^2. \end{aligned}$$

Now, using properties of $\langle\, ,\rangle$, we will prove the *Schwarz Inequality.* This will then give the triangle inequality for the metric d.

Proposition 2 (Schwarz Inequality) *For any* $x,y \in \mathbf{R}^n$, *we have*

$$\langle x,y\rangle^2 \leq \langle x,x\rangle\langle y,y\rangle. \tag{4}$$

Proof: By property (iv) of $\langle\ ,\ \rangle$, for any real number t we have

$$\langle x+ty, x+ty\rangle \geq 0.$$

Using bilinearity and symmetry, this is

$$\langle x,x\rangle + 2\langle x,y\rangle t + \langle y,y\rangle t^2 \geq 0. \tag{5}$$

This quadratic polynomial in t with real coefficients can have only one minimum, so (5) shows it cannot have two distinct roots. It follows that the discriminant cannot be > 0. That is,

$$(2\langle x,y\rangle)^2 - 4\langle y,y\rangle\langle x,x\rangle \leq 0.$$

This is equivalent to (4) and the Schwarz Inequality is proved. ∎

Now take $x, y, z \in \mathbf{R}^n$. We claim that $d(x,y)+d(y,z) \geq d(x,z)$ holds.
Proof: We apply the Schwarz Inequality to $x-y$ and $y-z$ to get

$$\langle x-y, y-z\rangle \leq \sqrt{\langle x-y, x-y\rangle}\sqrt{\langle y-z, y-z\rangle} = d(x,y)d(y,z) \tag{6}$$

We will show that (6) is equivalent to

$$(d(x,y)+d(y,z))^2 \geq (d(x,z))^2. \tag{7}$$

Rewriting (7) as

$$\langle x-y, x-y\rangle + 2d(x,y)d(y,z) + \langle y-z, y-z\rangle \geq \langle x-z, x-z\rangle,$$

and then expanding out and cancelling some terms gives

$$-2\langle x,y\rangle + 2\langle y,y\rangle - 2\langle y,z\rangle + 2d(x,y)d(y,z) \geq -2\langle x,z\rangle$$

or

$$-\langle x,y\rangle + \langle y,y\rangle - \langle y,z\rangle + \langle x,z\rangle + d(x,y)d(y,z) \geq 0$$

or

$$d(x,y)d(y,z) \geq \langle x-y, y-z\rangle$$

which is (6). ∎

Bilinear Forms

Definition Let V be a real vector space. A *bilinear form* on V is a function $f : V \times V \to \mathbf{R}$ which satisfies

(i) $f(u+v, w) = f(u,w) + f(v,w)$,

(i′) $f(u, v+w) = f(u,v) + f(u,w)$, and

(iii) $f(\alpha u, v) = \alpha f(u, v) = f(u, \alpha v)$ for $\alpha \in \mathbf{R}$.

The bilinear form is *symmetric* if

$$f(u, v) = f(v, u) \quad \text{for all } \ u, v \in V \, ;$$

and the symmetric bilinear form f is *positive definite* if

$$f(v, v) > 0 \quad \text{for } \ v \neq 0 \, .$$

Exercise. Show that $f(v, 0) = 0 = f(0, v)$ for any $v \in V$.

Examples

(1) The inner product $\langle \, , \rangle$ on $\mathbf{R}^n$ is a positive definite symmetric bilinear form.

(2) Given $A = (a_{ij}) \in M_n(\mathbf{R})$, we define a bilinear form on $\mathbf{R}^n$ as follows:

$$\begin{aligned} f(x, y) &= xA({}^t y) = (x_1, \ldots, x_n) \begin{pmatrix} a_{11} & \ldots & a_{1n} \\ \vdots & & \\ a_{n1} & \ldots & a_{nn} \end{pmatrix} \begin{pmatrix} y_1 \\ \vdots \\ y_n \end{pmatrix} \\ &= \sum_{i,j=1}^{n} a_{ij} x_i y_j \, . \end{aligned}$$

The fact that f is bilinear follows from the fact that matrix multiplication distributes over matrix addition.

We get the inner product of example (1) by taking $A = I$.

Note that f is symmetric $\Leftrightarrow$ A is a symmetric matrix.

If we choose a basis $\alpha = \{v_1, \ldots, v_n\}$ for the finite-dimensional vector space V, then we can say a great deal about any bilinear form f on V. First we express $u, v \in V$ in terms of this basis ($u = x_1 v_1 + \cdots + x_n v_n, v = y_1 v_1 + \cdots + y_n v_n$) and then we use bilinearity of f to compute

$$f(u, v) = \sum_{i,j=1}^{n} x_i y_j f(v_i, v_j) \, .$$

If we set $a_{ij} = f(v_i, v_j)$, then we see that f is completely determined by the n^2 numbers $\{a_{ij}\}$. Letting A be the $n \times n$ matrix (a_{ij}), an equivalent way to write $f(u, v)$ is

$$f(u, v) = (x_1, \ldots, x_n) A \begin{pmatrix} y_1 \\ \vdots \\ y_n \end{pmatrix} .$$

So once we choose a basis for V, all bilinear forms on V can be written as in example (2). A is called the *matrix representation* of f with respect to the basis $\{v_1, \ldots, v_n\}$.

A natural question to ask at this point concerns our choice of basis; i.e., if we had picked a different basis $\beta = \{w_1, \ldots, w_n\}$ for V, how would that affect the matrix representation of f? We might expect the answer to be the same as for endomorphisms (namely, that $M_\beta(\phi) = CM_\alpha(\phi)C^{-1}$ for some nonsingular matrix C), but this is not quite right. If we let B be the matrix representation of f with respect to the basis $\{w_1, \ldots, w_n\}$, then there is a nonsingular matrix C such that

$$B = CA{}^tC\,.$$

A proof of this relationship is sketched in the exercises.

Exercises

(1) One can make $\mathbf{R}^n \times \mathbf{R}^n$ into a vector space by taking for (x, y) the $2n$-tuple $(x_1, \ldots, x_n, y_1, \ldots, y_n)$. Show that if this is done, then

$$\langle\,,\rangle : \mathbf{R}^n \times \mathbf{R}^n \to \mathbf{R}$$

is not a linear map.

(2) Show that $\langle x, y\rangle = 0$ for all x implies $y = 0$.

(3) Show that equality holds in the Schwarz Inequality exactly when x and y are linearly dependent; i.e., $\langle x, y\rangle^2 = \langle x, x\rangle\langle y, y\rangle \Leftrightarrow$ there is a real number r such that $x = ry$.

(4) Prove the following refinement of the Schwarz Inequality: for any $x = (x_1, \ldots, x_n)$, $y = (y_1, \ldots, y_n)$ in $\mathbf{R}^n$, we have

$$|\langle x, y\rangle| \le \sum_{i=1}^{n} |x_i y_i| \le \|x\|\,\|y\|\,.$$

(Hint: Observe that $2\alpha\beta \le \alpha^2 + \beta^2$ for any $\alpha, \beta \in \mathbf{R}$, and set

$$\alpha = \frac{x_i}{\|x\|}\,, \quad \beta = \frac{y_i}{\|y\|}\,.$$

Now sum these inequalities over i.)

(5) The *rank* of a bilinear form f is defined to be the rank of any matrix representation of f. Show that this definition makes sense (i.e., if $B = CA{}^tC$ (C nonsingular), then A and B have the same rank).

(6) Given a symmetric bilinear form f on the real vector space V, the *quadratic form* associated to f is the function

$$g : V \to \mathbf{R}$$

defined by $g(v) = f(v,v)$. Show that f and g are related by

$$f(u,v) = \tfrac{1}{4}(g(u+v) - g(u-v)) .$$

(7) If we define a quadratic form on $V = \mathbf{R}^2$ by

$$g(x_1, x_2) = x_1^2 + 3x_1x_2 - 2x_2^2 ,$$

find the matrix of its associated symmetric bilinear form.

(8) Show that the real matrix A represents a symmetric bilinear form $\Leftrightarrow$ A is a symmetric matrix.

(9) Prove the relationship $B = CA^tC$ (where A, B are matrix representations of the bilinear form f with respect to the bases $\{v_1, \ldots, v_n\}$, $\{w_1, \ldots, w_n\}$) by filling in the details of the following steps.

(i) Define $C = (c_{ij})$ by

$$w_i = \sum_{j=1}^{n} c_{ij} v_j$$

and notice that this is just the change-of-basis matrix occuring in the proof of Proposition 4 (Chapter II).

(ii) Express $u, v \in V$ as

$$\begin{aligned} u &= x_1v_1 + \cdots + x_nv_n &= x_1'w_1 + \cdots + x_n'w_n \\ v &= y_1v_1 + \cdots + y_nv_n &= y_1'w_1 + \cdots + y_n'w_n \end{aligned}$$

and show that $x = (x_1, \ldots, x_n)$ is equal to $x'C = (x_1', \ldots, x_n')C$. Similarly, $y = y'C$.

(iii) Use the result of (ii) to show that $f(u,v) = xA({}^ty) = x'(CA^tC)({}^ty')$. Conclude that $B = CA^tC$.

B. Real Inner Product Spaces, Normed Vector Spaces

Let V be a real vector space. $(V, \langle\,,\rangle)$ is called a *real inner product space* if $\langle\,,\rangle$ is a positive definite symmetric bilinear form on V. As in section A, we define a norm on V by $\|v\| = \sqrt{\langle v, v\rangle}$. Note that $\| \quad \|^2$ is just the quadratic form associated to $\langle\ ,\rangle$. If it is understood that the field in question is $\mathbf{R}$, we will simply refer to $(V, \langle\ ,\rangle)$ as an inner product space.

It is worth observing that our proof of the Schwarz Inequality holds for any inner product space, finite-dimensional or not.

Example

If V is finite-dimensional, it is easy to define an inner product on V. Take any basis $\{v_1, \ldots, v_n\}$ for V, define $\langle v_i, v_j\rangle = \delta_{ij}$, and extend bilinearly. If $x = x_1v_1 + \cdots + x_nv_n$, $y = y_1v_1 + \cdots + y_nv_n$, then we get $\langle x, y\rangle = x_1y_1 + \cdots + x_ny_n$. This is clearly positive definite, bilinear, and symmetric.

Example

We will now examine an infinite-dimensional inner product space in some detail. Let V be the set of all real sequences $x = \{x_i\}$ such that Σx_i^2 exists; i.e., V consists of all square-summable real sequences. We add sequences and multiply by scalars in the obvious way; to see that the sum of two square-summable sequences is square-summable, we define

$$\langle x, y\rangle = \sum_{i=1}^{\infty} x_iy_i\ ,$$

and show that this series converges absolutely. By the refinement of the Schwarz Inequality occuring in the exercises of section A, we have

$$|x_1y_1| + \cdots + |x_ny_n| \le \sqrt{\sum_{i=1}^{n} x_i^2}\ \sqrt{\sum_{i=1}^{n} y_i^2} \le \sqrt{\sum_{i=1}^{\infty} x_i^2}\ \sqrt{\sum_{i=1}^{\infty} y_i^2}$$

which holds for all n. But then the monotonic sequence of partial sums $s_n = |x_1y_1| + \cdots + |x_ny_n|$ is bounded and so converges.

Now suppose that $\{x_i\}$ and $\{y_i\}$ are square-summable. Then

$$\begin{aligned} \sum(x_i + y_i)^2 &= \sum(x_i^2 + 2x_iy_i + y_i^2) \\ &= (\sum x_i^2) + (\sum y_i^2) + 2\langle x, y\rangle\ , \end{aligned}$$

and so $\{x_i\} + \{y_i\}$ is also square-summable. It is trivial to see that if $r \in \mathbf{R}$ and $\{x_i\} \in V$, then $\{rx_i\}$ is square-summable. Hence V is a real vector space.

Exercise. Show that $\langle\ ,\rangle$ defined above for square-summable sequences is positive definite, bilinear, and symmetric.

Thus we have an infinite-dimensional vector space with an inner product on it. It is called *ℓ_2-space* or *Hilbert Space*.

Definition Let $(V, \langle\, , \rangle)$ be an inner product space. We say that two vectors $u, v \in V$ are *orthogonal* if $\langle u, v\rangle = 0$. A set S of vectors in V is an *orthnormal set* if each $v \in S$ is a *unit vector* (i.e., $\|v\| = 1$) and any two vectors in S are orthogonal.

Example

The standard basis for $\mathbf{R}^n$ is an orthonormal set by (v) of Proposition 1.

Proposition 3 *Any nontrivial finite-dimensional inner product space* $(V, \langle\, , \rangle)$ *has an orthonormal basis.*

Proof: First, we choose any basis $\{v_1, \ldots, v_n\}$ for V, and then we set $v_1' = v_1$ and

$$v_2' = v_2 - \frac{\langle v_2, v_1'\rangle}{\langle v_1', v_1'\rangle}\, v_1' \,.$$

Note that v_1' and v_2' are orthogonal since

$$\begin{aligned}\langle v_1', v_2'\rangle &= \langle v_1', v_2 - \tfrac{\langle v_2, v_1'\rangle}{\langle v_1', v_1'\rangle}\, v_1'\rangle \\ &= \langle v_1', v_2\rangle - \tfrac{\langle v_2, v_1'\rangle}{\langle v_1', v_1'\rangle}\, \langle v_1', v_1'\rangle = 0 \,.\end{aligned}$$

Also note that v_1', v_2' are nonzero (since $\{v_1, v_2\}$ is linearly independent) and v_1, v_2 are in the span of $\{v_1', v_2'\}$.

More generally, define

$$v_m' = v_m - \frac{\langle v_m, v_{m-1}'\rangle}{\langle v_{m-1}', v_{m-1}'\rangle} v_{m-1}' - \cdots - \frac{\langle v_m, v_1'\rangle}{\langle v_1', v_1'\rangle} v_1'$$

and note that v_m' is orthogonal to $v_1', \ldots, v_{m-1}'$. Furthermore, v_m' is not the zero vector and $v_m \in \operatorname{Span}(v_1', \ldots, v_m')$. If we let

$$\begin{aligned} u_1 &= \tfrac{1}{\|v_1'\|}\, v_1' \,, \\ &\vdots \\ u_n &= \tfrac{1}{\|v_n'\|}\, v_n' \,, \end{aligned}$$

then $\{u_1, \ldots, u_n\}$ is an orthnormal set of vectors which span V. This is our desired basis. ∎

The technique used in the preceding proof to generate the orthonormal basis $\{u_1, \ldots, u_n\}$ is called the *Gram-Schmidt orthonormalization process.*

Definition Let $(V, \langle\, , \rangle)$ be an inner product space. A linear map $\phi : V \to V$ is said to be *orthogonal* if for all $u, v \in V$ we have

$$\langle \phi(u), \phi(v)\rangle = \langle u, v\rangle \,.$$

Equivalently (when V is finite-dimensional), $A \in M_n(\mathbf{R})$ is an *orthogonal matrix* if

$$\langle xA, yA\rangle = \langle x, y\rangle$$

for all $x, y \in \mathbf{R}^n$. We denote the set of $n \times n$ orthogonal matrices by $O(n)$.

Clearly an orthogonal map preserves lengths (= norms) of vectors, because

$$\|\phi(v)\|^2 = \langle\phi(v), \phi(v)\rangle = \langle v, v\rangle = \|v\|^2.$$

Strangely enough, the converse is true.

Proposition 4 *If $\phi : V \to V$ preserves lengths, then it is orthogonal.*

Proof: Our hypothesis is:

(a) $\langle\phi(v), \phi(v)\rangle = \langle v, v\rangle$ for all v.

Our conclusion is:

(b) $\langle\phi(u), \phi(v)\rangle = \langle u, v\rangle$ for all u, v.

Statement (b) appears to be more general. But we will prove that (a) $\Rightarrow$ (b) by "polarizing the identity (a)" (a procedure we will also use in our study of normed algebras).

We apply (a) to the vector $u + v$:

$$\begin{aligned}\langle\phi(u+v), \phi(u+v)\rangle &= \langle u+v, u+v\rangle = \langle u, v\rangle + 2\langle u, v\rangle + \langle v, v\rangle \\ &= \langle\phi(u), \phi(u)\rangle + 2\langle u, v\rangle + \langle\phi(v), \phi(v)\rangle,\end{aligned}$$

and note that also

$$\begin{aligned}\langle\phi(u+v), \phi(u+v)\rangle &= \langle\phi(u) + \phi(v), \phi(u) + \phi(v)\rangle \\ &= \langle\phi(u), \phi(u)\rangle + 2\langle\phi(u), \phi(v)\rangle + \langle\phi(v), \phi(v)\rangle\end{aligned}$$

proving (b). ∎

We have discussed two different concepts of orthogonality (namely, orthogonal maps and orthogonal vectors). These notions are quite closely related.

Given the matrix $A \in M_n(\mathbf{R})$, note that $R_i = e_i A \in \mathbf{R}^n$ is the ith row of A. If A is orthogonal, then $\langle R_i, R_j\rangle = \langle e_i A, e_j A\rangle = \langle e_i, e_j\rangle = \delta_{ij}$. Said differently, the rows of A form an orthonormal basis for $\mathbf{R}^n$. (Exercise. Show that if $S \subseteq \mathbf{R}^n$ is any orthonormal set, then S is linearly independent.)

The converse is also true (namely, that a matrix whose rows form an orthonormal basis for $\mathbf{R}^n$ is orthogonal), but a proof of this fact will have to wait until we know more about orthogonal matrices.

Normed Vector Spaces

We saw in section A that an inner produce $\langle\,,\rangle$ on $\mathbf{R}^n$ defines a norm by

$$\|x\| = \sqrt{\langle x, x\rangle}\,. \tag{8}$$

Similarly, given the inner product space $(V, \langle\,,\rangle)$, we can define a norm on V by (8). Notice that this norm inherits some properties from $\langle\,,\rangle$. For example, for $r \in \mathbf{R}$ and $v \in V$, we have

$$\|rv\| = \sqrt{\langle rv, rv\rangle} = \sqrt{r^2}\,\sqrt{\langle v, v\rangle} = |r|\,\|v\|\,. \tag{9}$$

Also, positive definiteness of $\langle\,,\rangle$ tells us that

$$\|v\| \geq 0\,, \quad \text{and} \quad \|v\| = 0 \Leftrightarrow v = 0\,. \tag{10}$$

Finally, the Schwarz Inequality gives

$$\|u + v\| \leq \|u\| + \|v\| \tag{11}$$

for any $u, v \in V$. (To see this, square both sides to get

$$\begin{aligned} \|u+v\|^2 &= \langle u+v, u+v\rangle = \|u\|^2 + \|v\|^2 + 2\langle u, v\rangle \quad \text{and} \\ (\,\|u\| + \|v\|)^2 &= \|u\|^2 + \|v\|^2 + 2\|u\|\,\|v\|\,, \end{aligned}$$

and note that the Schwarz Inequality is just $|\langle u, v\rangle| \leq \|u\|\,\|v\|$.)

Definition Let V be a real vector space and suppose there is a function $\|\;\| : V \to \mathbf{R}$ satisfying properties (9), (10), and (11). Then we call $(V, \|\;\|)$ a *normed vector space.*

Note that $\|\;\|$ is a continuous function on V; so if we have a sequence of vectors $\{v_n\}$ converging to the vector v, then the sequence of numbers $\{\|v_n\|\}$ converges to $\|v\|$.

We have shown that every inner product space yields a normed space by using (8). Conversely, given a norm on V, can we define an inner product which will give us back our norm (via (8))? Unfortunately, the answer is "no" in general; however, we do have the following

Lemma Let $(V, \|\;\|)$ be a normed vector space. Then $\|\;\|$ comes from an inner product $\Leftrightarrow$ the norm satisfies the *parallelogram law*

$$\|u+v\|^2 - \|u-v\|^2 = 2(\,\|u\|^2 + \|v\|^2) \quad \text{for all} \;\; u, v \in V\,.$$

Proof: $\Rightarrow$ Exercise.

$\Leftarrow$ If there is an inner product which induces $\|\;\|$, then it must satisfy the *polar form*

$$\langle u, v\rangle = \tfrac{1}{4}(\,\|u+v\|^2 - \|u-v\|^2)\,. \tag{12}$$

(This is just exercise (6) from the previous section.)

With this in mind, we define a function on $V \times V$ by (12) and show that this is, indeed, an inner product which induces $\|\;\|$.

It is clear that $\langle\,,\rangle$ is symmetric, positive definite, and that it gives back $\|\;\|$ since

$$\langle v, v\rangle = \tfrac{1}{4}(\,\|v+v\|^2 - \|0\|^2) = \tfrac{1}{4}(2^2\|v\|^2) = \|v\|^2\,.$$

Now we show that

$$(13) \qquad \langle u+v, w\rangle = \langle u, w\rangle + \langle v, w\rangle$$

(and hence we get $\langle u, v+w\rangle = \langle u, v\rangle + \langle u, w\rangle$ by symmetry). Expanding the right hand side of (13) by using (12) gives

$$\begin{aligned}
\langle u, w\rangle + \langle v, w\rangle &= \tfrac{1}{4}(\|u+w\|^2 - \|u-w\|^2) + \tfrac{1}{4}(\|v+w\|^2 - \|v-w\|^2) \\
&= \tfrac{1}{4}(\|u+w\|^2 + \|v+w\|^2) - \tfrac{1}{4}(\|u-w\|^2 + \|v-w\|^2) \\
&= \tfrac{1}{8}(\|u+w+v+w\|^2 - \|u+w-v-w\|^2) \\
&\quad - \tfrac{1}{8}(\|u-w+v-w\|^2 - \|u-w-v+w\|^2) \\
&= \tfrac{1}{8}(\|u+v+2w\|^2 - \|u+v-2w\|^2) \\
\langle u, w\rangle + \langle v, w\rangle &= \tfrac{1}{2}\langle u+v, 2w\rangle
\end{aligned}$$

where the third equality holds by the parallelogram law, and the last holds by (12). Setting $v = 0$ in this last equation yields

$$(14) \qquad \langle u, w\rangle = \tfrac{1}{2}\langle u, 2w\rangle\ ,$$

and substituting $u+v$ for u in (14) gives (13).

To show bilinearity of $\langle\, ,\rangle$, we still must show that

$$(15) \qquad \langle ru, v\rangle = r\langle u, v\rangle \quad \text{for any} \quad r \in \mathbf{R}\,.$$

First suppose $r \in \mathbf{N} = \{1, 2, \ldots\}$ and observe that

$$\langle ru, v\rangle = \langle \underbrace{u + \cdots + u}_{r}, v\rangle = \underbrace{\langle u, v\rangle + \cdots + \langle u, v\rangle}_{r} = r\langle u, v\rangle\ ,$$

by using (13) repeatedly. Furthermore, note that

$$0 = \langle u + (-u), v\rangle = \langle u, v\rangle + \langle -u, v\rangle$$

so that $\langle -u, v\rangle = -\langle u, v\rangle$. Hence (15) holds when r is an integer.

Now suppose that $r = \frac{p}{q}$ is any rational number (p, q integers with $q \neq 0$). Then

$$\langle \tfrac{p}{q}u, v\rangle = p\langle \tfrac{1}{q}u, v\rangle = \left(\tfrac{p}{q}\right) q\langle \tfrac{1}{q}u, v\rangle = \tfrac{p}{q}\langle q(\tfrac{1}{q}u), v\rangle = \tfrac{p}{q}\langle u, v\rangle$$

and so (15) holds for the rationals.

Finally, choose any $r \in \mathbf{R}$ and let $\{r_i\}$ be a sequence of rational numbers converging to r. (We know such a sequence exists since the rationals are *dense* in the reals.) Note that $\{r_i u\}$ is a sequence of vectors converging to ru. Then we have

$$\begin{aligned}\langle ru, v\rangle &= \langle \lim_{i\to\infty}(r_i u), v\rangle = \lim_{i\to\infty}\langle r_i u, v\rangle \\ &= (\lim_{i\to\infty} r_i)\langle u, v\rangle = r\langle u, v\rangle\,.\end{aligned}$$

Note that we can "pull the limit out" of $\langle\,,\rangle$ since $\|\ \|$ is continuous. ∎

Now we consider an example of a normed vector space $(V, \|\ \|)$ in which $\|\ \|$ does not come from any inner product on V.

Example

Let $V = \mathbf{R}^2$ and define $\|\ \|$ by

$$\|(x_1, x_2)\| = |x_1| + |x_2|\,.$$

It is left as an exercise to show that this is actually a norm on $\mathbf{R}^2$.

If we let $x = (1,0)$ and $y = (0,1)$, then we see that $\|x+y\|^2 - \|x-y\|^2 = (|1| + |1|)^2 - (|1| + |-1|)^2 = 0$ whereas $2(\|x\|^2 + \|y\|^2) = 2(|1| + |1|\,) = 4$. So this norm does not satisfy the parallelogram law.

Exercises

(1) Given any set of vectors W from the inner product space $(V, \langle,\rangle)$, we define the *orthogonal complement* of W to be all vectors in V which are orthogonal to every $w \in W$; i.e., $W^\perp = \{v \in V \mid \langle v, w\rangle = 0 \text{ for all } w \in W\}$. Show that $W^\perp$ is a subspace of V. If V is finite-dimensional and if W is a subspace of V, show that $V \cong W \oplus W^\perp$.

(2) Let W be the subspace of $\mathbf{R}^3$ spanned by $u = (1,1,0)$ and $v = (1,0,-2)$. Find orthonormal bases for W and $W^\perp$. Verify that the union of these two bases is an orthonormal basis for $\mathbf{R}^3$.

(3) Let ϕ be an orthogonal endomorphism on the finite-dimensional inner product space $(V, \langle\,,\rangle)$, and suppose that W is a ϕ-stable subspace of V. Show that

 (i) $\phi|_W$ is orthogonal, and

 (ii) $W^\perp$ is ϕ-stable. (Hint: ϕ is nonsingular — a fact we will prove later — so $\phi(W) = W$.)

(4) Show that $\|\ \| : \mathbf{R}^2 \to \mathbf{R}$ defined by $\|(x_1, x_2)\| = |x_1| + |x_2|$ is a norm.

(5) Draw a picture in $\mathbf{R}^2$ to justify the name "parallelogram law."

(6) For $V = \mathbf{R}^n$, show that the following define norms on $\mathbf{R}^n$.

(i) $\|(x_1, \ldots, x_n)\|_p = (|x_1|^p + \cdots + |x_n|^p)^{\frac{1}{p}}$, where $1 \leq p < \infty$.

(ii) $\|(x_1, \ldots, x_n)\|_\infty = \max\{|x_1|, \ldots, |x_n|\}$.

Can you decide which of these norms comes from an inner product?

(7) Let V be the infinite-dimensional vector space of real-valued continuous functions on $[-\pi, \pi]$. Show that

$$\langle f, g \rangle = \int_{-\pi}^{\pi} f(t)g(t)dt$$

defines an inner product on V. What is $\|f\|$ if $f(t) = t + 3$? Show that $\{1, \sin t, \sin(2t), \ldots, \cos t, \cos(2t), \ldots\}$ is an orthogonal subset of V.

C. Complex Inner Product Spaces

A complex vector space is simply a vector space whose scalar field is $\mathbf{C}$. An inner product on a complex vector space is much the same as the real inner products we have already studied; however, there is one important difference from the real case which arises because conjugation is a nontrivial operation on $\mathbf{C}$.

First we define a *Hermitian inner product* $\langle\, , \rangle$ on $\mathbf{C}^n$: for $x = (\alpha_1, \ldots, \alpha_n), y = (\beta_1, \ldots, \beta_n) \in \mathbf{C}^n$ we let

$$\langle x, y \rangle = \alpha_1 \bar{\beta}_1 + \cdots + \alpha_n \bar{\beta}_n.$$

Proposition 5 *$\langle\, , \rangle$ has the following properties.*

(i) $\langle\, , \rangle$ *is* Hermitian symmetric; *i.e.*, $\langle x, y \rangle = \overline{\langle y, x \rangle}$,

(ii) $\langle x + y, z \rangle = \langle x, z \rangle + \langle y, z \rangle$,

(ii′) $\langle x, y + z \rangle = \langle x, y \rangle + \langle x, z \rangle$,

(iii) $\langle \alpha x, y \rangle = \alpha \langle x, y \rangle$,

(iii′) $\langle x, \alpha y \rangle = \bar{\alpha} \langle x, y \rangle$,

(iv) $\overline{\langle x, y \rangle} = \langle \bar{x}, \bar{y} \rangle$,

(v) $\langle\, , \rangle$ *is a positive definite; i.e., $\langle x, x \rangle$ is a real number ≥ 0, and $\langle x, x \rangle = 0 \Leftrightarrow x = (0, \ldots, 0)$,*

(vi) $\langle e_i, e_j \rangle = \delta_{ij}$ *where* $\{e_1, \ldots, e_n\}$ *is the standard basis for* $\mathbf{C}^n$, *and*

(vii) $\langle \, , \rangle$ *is nondegenerate.*

Proof: Exercise. (Hint for (v): Recall that if $\alpha = a + bi \in \mathbf{C}$, then $\alpha\bar{\alpha} = a^2 + b^2 \in \mathbf{R}$.) ∎

Properties (ii) and (iii) say together that $\langle , \rangle$ is linear in the first variable, while (ii′) and (iii′) say that $\langle \, , \rangle$ is *conjugate linear* in the second variable. So our inner product on $\mathbf{C}^n$ is said to be *Hermitian linear.*

We could have defined our inner product on $\mathbf{C}^n$ by formula (1), and then it would still be bilinear. Unfortunately, our inner product would no longer be positive definite because $\alpha\alpha$ is not always a real number if $\alpha \in \mathbf{C}$. Since positive definiteness is crucial for defining norms, we are willing to sacrifice bilinearity.

Definition Let V be a complex vector space. A *Hermitian form* on V is a function $f : V \times V \to \mathbf{C}$ which is Hermitian symmetric and Hermitian linear; i.e.,

$$f(u, v) = \overline{f(v, u)}, \quad \text{and}$$
$$f(\alpha u + \beta v, w) = \alpha f(u, w) + \beta f(v, w)$$

where $u, v, w \in V$ and $\alpha, \beta \in \mathbf{C}$.

If V is a complex vector space, then $(V, \langle \, , \rangle)$ is called a *complex inner product space* if $\langle \, , \rangle$ is a positive definite Hermitian form. The Schwarz Inequality still holds in complex inner product spaces, but we need to modify its statement and proof for the present case.

Proposition 6 (Schwarz Inequality) *For any* u, v *in the complex inner product space* $(V, \langle \, , \rangle)$, *we have*

$$|\langle u, v \rangle|^2 \leq \langle u, u \rangle \langle v, v \rangle .$$

Proof: We will use the fact that $\alpha\bar{\alpha} = |\alpha|^2$.

First, observe that if $\langle u, v \rangle$ is real, then our previous proof works without any problems.

If $\alpha = \langle u, v \rangle$ is not real, then we have

$$\langle \frac{u}{\alpha}, v \rangle = \frac{1}{\alpha} \langle u, v \rangle = \frac{\langle u, v \rangle}{\langle u, v \rangle} = 1$$

so that $\langle \frac{u}{\alpha}, v \rangle$ is real. Thus we have

$$1 = \langle \frac{u}{\alpha}, v \rangle^2 \leq \langle \frac{u}{\alpha}, \frac{u}{\alpha} \rangle \langle v, v \rangle = \frac{1}{\alpha\bar{\alpha}} \langle u, u \rangle \langle v, v \rangle$$

so that $\alpha\bar{\alpha} = \langle u, u \rangle \langle v, v \rangle$. Since $\alpha = \langle u, v \rangle$, we have our result. ∎

Given two vectors u, v from the complex inner product space $(V, \langle\, ,\rangle)$, we say they are *orthogonal* if $\langle u, v\rangle = 0$ (just as we did in the real case). The Gram-Schmidt process works exactly the same, too, so every finite-dimensional complex inner product space has an orthonormal basis.

We define a *norm* on V by

$$\|v\| = \sqrt{\langle v, v\rangle}\,. \tag{16}$$

It is left as an exercise to show that this norm inherits the same properties from $\langle\, ,\rangle$ as in the real case; i.e.,

$$\|\alpha v\| = |\alpha|\,\|v\| \quad \text{for} \quad \alpha \in \mathbf{C}\,,\ v \in V\,, \tag{17}$$

$$\|v\| \geq 0\,, \quad \text{and} \quad \|v\| = 0 \Leftrightarrow v = 0\,, \quad \text{and} \tag{18}$$

$$\|u + v\| \leq \|u\| + \|v\|\,. \tag{19}$$

Hence the definition of a (complex) normed vector space is exactly the same as before. Furthermore, it is still true that a norm on V comes from an inner product (via (16)) $\Leftrightarrow$ the norm satisfies the parallelogram law.

Definition Let $(V, \langle\, ,\rangle)$ be a complex inner product space. A linear map $\phi : V \to V$ is *unitary* if for any $u, v \in V$ we have

$$\langle \phi(u), \phi(v)\rangle = \langle u, v\rangle.$$

Equivalently (when V is finite-dimensional), $A \in M_n(\mathbf{C})$ is a *unitary matrix* if

$$\langle xA, yA\rangle = \langle x, y\rangle$$

for all $x, y \in \mathbf{C}^n$. We denote the set of $n \times n$ unitary matrices by $U(n)$.

Proposition 7 *$U(n)$ is a group under matrix multiplication.*

Proof: Matrix multiplication is associative, and clearly the identity matrix is unitary. If $A, B \in U(n)$, then $\langle xAB, yAB\rangle = \langle xA, yA\rangle = \langle x, y\rangle$, showing that AB is also unitary.

Note that if $A = (a_{ij}) \in M_n(\mathbf{C})$, then we define the *conjugate* of A to be

$$\bar{A} = (\bar{a}_{ij})\,.$$

Since the operations of conjugation and taking transposes commute, the symbol ${}^t\bar{A}$ is unambiguous. It remains to prove that if A is unitary, then A is an isomorphism and A^{-1} is also unitary. For this, and other purposes, the following lemma is important.

Lemma 1 *For $A \in M_n(\mathbf{C})$ and $x, y \in \mathbf{C}^n$ we have*

$$\langle xA, y\rangle = \langle x, y\,{}^t\bar{A}\rangle\,.$$

Proof: Let $A = (a_{ij})$ and calculate

$$xA = (x_1a_{11} + \cdots + x_na_{n1}, \ldots, x_1a_{1n} + \cdots + x_na_{nn})$$
$$y\,{}^t\bar{A} = (y_1\bar{a}_{11} + \cdots + y_n\bar{a}_{1n}, \ldots, y_1\bar{a}_{n1} + \cdots + y_n\bar{a}_{nn})\,.$$

Thus we see that $\langle xA, y\rangle$ is equal to

$$(x_1a_{11} + \cdots + x_na_{n1})\bar{y}_1 + \cdots + (x_1a_{1n} + \cdots + x_na_{nn})\bar{y}_n\,,$$

whereas the $\langle x, y\,{}^t\bar{A}\rangle$ is

$$x_1(a_{11}\bar{y}_1 + \cdots + a_{1n}\bar{y}_n) + \cdots + x_n(a_{n1}\bar{y}_1 + \cdots + a_{nn}\bar{y}_n).$$

Inspection shows these two expressions contain exactly the same terms. ∎

Corollary For $A \in M_n(\mathbf{R})$ and $x, y \in \mathbf{R}^n$, Lemma 1 gives $\langle xA, y\rangle = \langle x, y\,{}^tA\rangle$.

Continuing the proof of Proposition 7, we use Lemma 1 to show that

$$A \in M_n(\mathbf{C}) \text{ is unitary } \Leftrightarrow {}^t\bar{A} = A^{-1}.$$

By Lemma 1, $\langle e_iA, e_jA\rangle = \langle e_i, (e_jA)\,{}^t\bar{A}\rangle$. Thus

$$\langle e_iA, e_jA\rangle = \langle e_i, e_j\rangle = \delta_{ij} \Leftrightarrow A\,{}^t\bar{A} = I.$$

So if A is unitary, then A^{-1} exists and is equal to ${}^t\bar{A}$. Furthermore, $\langle xA^{-1}, yA^{-1}\rangle = \langle xA^{-1}A, yA^{-1}A\rangle = \langle x, y\rangle$. This shows A^{-1} is also unitary and completes the proof that $U(n)$ is a group. ∎

In the course of proving Proposition 7, we also showed that the rows of a unitary matrix form an orthonormal basis for $\mathbf{C}^n$.

Next we want to prove a result for unitary matrices like we did for orthogonal matrices.

Proposition 8 *If $A \in M_n(\mathbf{C})$ and A preserves lengths, then A is unitary.*

Proof: To prove this theorem, the following lemma will be useful.

Lemma 2 *If A preserves lengths, then*

$$\langle e_iA, e_jA\rangle = -\langle e_jA, e_iA\rangle \quad \textit{for } i \neq j\,.$$

Proof:

$$\langle (e_i + e_j)A, (e_i + e_j)A\rangle = \langle e_i + e_j, e_i + e_j\rangle = \langle e_i, e_i\rangle + \langle e_j, e_j\rangle = 2.$$

But the left hand side also equals

$$\begin{aligned}\langle e_iA+e_jA, e_iA+e_jA\rangle &= \langle e_iA, e_iA\rangle + \langle e_iA, e_jA\rangle + \langle e_jA, e_iA\rangle + \langle e_jA, e_jA\rangle \\ &= 2 + \langle e_iA, e_jA\rangle + \langle e_jA, e_iA\rangle .\end{aligned}$$

This proves Lemma 2. ∎

Now we can prove Proposition 8. For $i \neq j$, let

$$x = \alpha e_i + \beta e_j$$

where $\alpha, \beta \in \mathbf{C}$ are not yet specified. Then

$$\langle x, x\rangle = \langle \alpha e_i + \beta e_j, \alpha e_i + \beta e_j\rangle = \alpha\bar{\alpha} + 0 + 0 + \beta\bar{\beta}. \tag{20}$$

But, by Lemma 2, we also have

$$\begin{aligned}\langle x, x\rangle &= \langle xA, xA\rangle = \langle \alpha e_iA + \beta e_jA, \alpha e_iA + \beta e_jA\rangle \\ &= \alpha\bar{\alpha}\langle e_iA, e_iA\rangle + \alpha\bar{\beta}\langle e_iA, e_jA\rangle + \beta\bar{\alpha}\langle e_jA, e_iA\rangle \\ &\qquad + \beta\bar{\beta}\langle e_jA, e_jA\rangle \\ &= \alpha\bar{\alpha} + \beta\bar{\beta} + (\alpha\bar{\beta} - \beta\bar{\alpha})\langle e_iA, e_jA\rangle .\end{aligned} \tag{21}$$

Comparing (20) and (21) gives

$$(\alpha\bar{\beta} - \beta\bar{\alpha})\langle e_iA, e_jA\rangle = 0.$$

Letting $\alpha = i$ and $\beta = 1$ gives $\alpha\bar{\beta} - \beta\bar{\alpha} = 2i$, and hence

$$\langle e_iA, e_jA\rangle = 0 \quad \text{for} \quad i \neq j .$$

So for $x = x_1e_1 + \cdots + x_ne_n$, $y = y_1e_1 + \cdots + y_ne_n$ we have

$$\begin{aligned}\langle xA, yA\rangle &= \langle (x_1e_1 + \cdots + x_ne_n)A, (y_1e_1 + \cdots + y_ne_n)A\rangle \\ &= x_1\bar{y}_1 + \cdots + x_n\bar{y}_n = \langle x, y\rangle ,\end{aligned}$$

proving that A is unitary. ∎

Exercises

(1) Show that the norm defined by (16) inherits properties (17)–(19) from the Hermitian inner product $\langle\ ,\ \rangle$.

(2) Prove that a norm on a complex vector space comes from a Hermitian inner product (via (16)) $\Leftrightarrow$ the norm satisfies the parallelogram law. Give an example of a norm on $\mathbf{C}^2$ which does not come from any inner product.

(3) Let V be the complex vector space of all continuous complex-valued functions on $[0,1]$. Show that

$$\langle f, g\rangle = \int_0^1 f(t)\overline{g(t)}dt$$

defines an inner product on V. What is the norm of $e^{it} = \cos t + i\sin t$?

(4) Given $A \in M_n(\mathbf{C})$, show that

$$^t(\bar{A}) = \overline{(^tA)}\,.$$

(And so $^t\bar{A}$ is unambiguous.)

(5) Let $(V, \langle\ ,\rangle)$ be a complex inner product space, and suppose $\phi \in \text{End}(V)$ is unitary. If W is a ϕ-stable subspace of V, show that $W^\perp$ is also ϕ-stable.

(6) The matrix $A \in M_n(\mathbf{C})$ is called *Hermitian* if

$$^t\bar{A} = A\,.$$

(Note that the set of Hermitian matrices contains the set of symmetric matrices as a subset.) Show that the eigenvalues of a Hermitian matrix are real numbers. (Hint: Consider $\langle xA, x\rangle$ where x is an eigenvector for A.)

(7) The matrix $A \in M_n(\mathbf{C})$ is called *skew-Hermitian* if

$$^t\bar{A} = -A\,.$$

Show that the eigenvalues of a skew -Hermitian matrix are purely imaginary. (We say $\alpha \in \mathbf{C}$ is purely imaginary if $\alpha = bi$ where $b \in \mathbf{R}$.)

(8) The matrix $A \in M_n(\mathbf{C})$ is called *positive definite* if there is a nonsingular matrix $B \in M_n(\mathbf{C})$ such that

$$A = B^t\bar{B}\,.$$

Show that the eigenvalues of a positive definite matrix are real and positive. (Hint: Consider $\langle xB, xB\rangle$ where x is an eigenvector for A.) Note that, in particular, this holds for $A \in M_n(\mathbf{R})$.

(9) Given the Hermitian matrix $A \in M_n(\mathbf{C})$, show that the function $f : \mathbf{C}^n \times \mathbf{C}^n \to \mathbf{C}$ defined by

$$f(x, y) = xA(\overline{^ty})$$

is a Hermitian form on $\mathbf{C}^n$. Conversely, given the basis $\{v_1, \ldots, v_n\}$ for the complex vector space V, and given the Hermitian form f on V, we get a matrix $A = (a_{ij})$ associated to f as above (i.e., $a_{ij} = f(v_i, v_j)$). Show that A is a Hermitian matrix. How does A change if we choose a different basis for V?

For exercises (10)–(14), let $(V, \langle\ ,\rangle)$ be an inner product space (real or complex).

(10) Show that each $v \in V$ determines a functional $\phi_v \in V^*$ defined by

$$\phi_v(w) = \langle w, v\rangle\ .$$

However, ψ_v defined by $\psi_v(w) = \langle v, w\rangle$ is not linear unless V is a real vector space.

(11) If V is finite-dimensional and if $\phi \in V^*$ is any functional, show that there is a unique $v \in V$ such that $\phi(w) = \langle w, v\rangle$ for all $w \in V$.

(12) Suppose V is finite-dimensional and let $\psi \in \mathrm{End}(V)$. Use exercise (11) to show that there is a unique endomorphism ψ^* of V such that

$$\langle \psi(u), v\rangle = \langle u, \psi^*(v)\rangle$$

for all $u, v \in V$. We call ψ^* the *adjoint* of ψ. (Hint: Fix $v \in V$ so that ϕ defined by $\phi(u) = \langle \psi(u), v\rangle$ is a functional on V. Then there is a unique $v' \in V$ such that $\phi(u) = \langle u, v'\rangle$ for all $u \in V$. Define ψ^* by

$$\psi^*(v) = v'\ .)$$

(13) Continuing exercise (12), choose an orthonormal basis α for V. Show that if $A = M_\alpha(\psi)$, then ${}^t\bar{A} = M_\alpha(\psi^*)$.

(14) Show that $A \in M_n(\mathbf{C})$ represents a unitary map $\phi \in \mathrm{End}(V)$ (with respect to an orthonormal basis of V) $\Leftrightarrow$ A is a unitary matrix. State and prove an analogous result for orthogonal maps/matrices.

(Thus we see that if ϕ is orthogonal or unitary, then it is nonsingular.)

D. Orthogonal and Unitary Groups

Recall that $O(n)$ is the set of $n \times n$ orthogonal matrices. Because of the corollary to Lemma 1, our proof that $U(n)$ is a group (whose operation is matrix multiplication) carries over almost verbatim to show that $O(n)$ is also a multiplicative group. In fact, we can think of $O(n)$ as a subgroup of $U(n)$ where the inclusion is induced by the natural inclusion of $\mathbf{R}$ into $\mathbf{C}$.

Proposition 9 *For $A \in M_n(\mathbf{R})$, the following statements are equivalent:*

(i) *A is orthogonal,*

(ii) $\langle e_i A, e_j A\rangle = \delta_{ij}$,

(iii) *A maps an orthonormal basis (of $\mathbf{R}^n$) to an orthonormal basis,*

(iv) *the rows of A form an orthonormal basis*

(v) *the columns of A form an orthonormal basis, and*

(vi) ${}^tA = A^{-1}$.

Proof: Exercise. ▮

In Proposition 9, if we make the obvious adjustments, then we get an analogous characterization of unitary matrices. From the equivalence of (iv) and (v), we conclude that A is orthogonal (unitary) $\Leftrightarrow$ ${}^tA$ is orthogonal (unitary).

Proposition 10 *If $A \in U(n)$, then $\det A$ is a complex number of unit ength. If $A \in O(n)$, then $\det A \in \{1, -1\}$.*

Proof: For A unitary, we know that $A^{-1} = {}^t\bar{A}$. Thus

$$1 = \det(I) = \det(AA^{-1}) = \det A \det({}^t\bar{A}) = \det A\, \overline{\det A}$$

so that $|\det A| = \sqrt{\det A\, \overline{\det A}} = 1$.

For A orthogonal, we know that A is also unitary (by the inclusion $O(n) \subseteq U(n)$) so that $|\det A| = 1$. But $\det A$ is a real number, so we must have $\det A = 1$ or $\det A = -1$. ▮

Definition The *special orthogonal group* is

$$SO(n) = \{A \in O(n) \mid \det A = 1\}$$

(also called the *rotation group*). Similarly, the *special unitary group* is

$$SU(n) = \{A \in U(n) \mid \det A = 1\}\,.$$

The fact that these are, indeed, groups is an easy exercise.

Definition We say that the matrix $A \in M_n(\mathbf{R})$ is *skew-symmetric* if

$${}^tA = -A\,.$$

The matrix $A \in M_n(\mathbf{C})$ is *skew-Hermitian* if

$$^t\bar{A} = -A\,.$$

The set of all $n \times n$ skew-symmetric matrices is denoted by $o(n)$, and the skew-Hermitian matrices are denoted by $u(n)$.

Note that $o(n)$ and $u(n)$ are not multiplicative groups. For example, if

$$A = \begin{pmatrix} 0 & 1 \\ -1 & 0 \end{pmatrix},$$

then A is skew-symmetric, but $A^2 = \begin{pmatrix} -1 & 0 \\ 0 & -1 \end{pmatrix}$ is not. However, $o(n)$ is a subspace of the n^2-dimensional real vector space $M_n(\mathbf{R})$. (Exercise. Show this.)

It is not true that $u(n)$ is a subspace of the complex vector space $M_n(\mathbf{C})$ (since a complex multiple of a skew-Hermitian matrix need not be skew-Hermitian); but if we think of $M_n(\mathbf{C})$ as a $(2n)^2$-dimensional real vector space, then $u(n)$ is a (real) subspace. For example, if

$$A = \begin{pmatrix} 2i & 1+i \\ -1+i & i \end{pmatrix},$$

then A is skew-Hermitian, and so is rA for any $r \in \mathbf{R}$, but

$$iA = \begin{pmatrix} -2 & -1+i \\ -1-i & -1 \end{pmatrix}$$

is not skew-Hermitian.

Proposition 11 *The dimension of* $o(n)$ *is* $\frac{n(n-1)}{2}$, *and the dimension of* $u(n)$ *is* n^2.

Proof: Exercise. (Hint: Bases for both spaces are easy to write down.) ∎

Recall the definition of a Lie Algebra from Section II B (i.e., an algebra whose multiplication is anticommutative and satisfies the Jacobi identity). We make the real vector spaces $o(n)$ and $u(n)$ into Lie Algebras by defining

$$[A, B] = AB - BA\,,$$

where AB denotes matrix multiplication. To see that $[A, B]$ is skew-symmetric whenever $A, B \in o(n)$, we observe that

$$\begin{aligned} {}^t[A,B] &= {}^t(AB - BA) = {}^t(AB) - {}^t(BA) \\ &= {}^tB\,{}^tA - {}^tA\,{}^tB = -({}^tA\,{}^tB - {}^tB\,{}^tA) \\ &= -((-A)(-B) - (-B)(-A)) = -[A, B]\,. \end{aligned}$$

Similarly, $[A, B] \in u(n)$ whenever $A, B \in u(n)$.

The groups $O(n), SO(n), U(n)$, and $SU(n)$ are examples of *Lie Groups.* (The definition of a Lie Group would take us too far afield, but we simply remark that they are important in various branches of mathematics.) Now $o(n)$ is the Lie Algebra associated to the Lie Group $O(n)$ (and also to $SO(n)$), and $u(n)$ is the Lie Algebra associated to $U(n)$. The Lie Algebra associated to $SU(n)$ is

$$su(n) = \{A \in u(n) \mid \text{trace } A = 0\} .$$

The idea of associating a Lie Algebra to a Lie Group is akin to associating f' to f in calculus. We do this because linear objects are easier to study than nonlinear ones, but we still get useful information about the associated nonlinear objects.

Exercises

(1) State and prove a proposition (analogous to Proposition 9) for unitary matrices.

(2) Show that if $A = (a_{ij})$ is an orthogonal or unitary matrix which is triangular, than A is diagonal.

(3) Show that $SO(n)$ and $SU(n)$ are, indeed, subgroups of $O(n)$ and $U(n)$, respectively.

(4) Show that $SO(2) = \left\{ \begin{pmatrix} \cos\theta & \sin\theta \\ -\sin\theta & \cos\theta \end{pmatrix} \middle| \theta \in \mathbf{R} \right\}$; also, show that $U(1) = \{e^{i\theta} = \cos\theta + i\sin\theta \mid \theta \in \mathbf{R}\}$. Now show that the map $\phi : SO(2) \to U(1)$ defined by $\phi\begin{pmatrix} \cos\theta & \sin\theta \\ -\sin\theta & \cos\theta \end{pmatrix} = e^{i\theta}$ is a group isomorphism (i.e., $\phi(AB) = \phi(A)\phi(B)$, and there is a map $\psi : U(1) \to SO(2)$ such that $\psi(e^{i\theta_1}e^{i\theta_2}) = \psi(e^{i\theta_1})\psi(e^{i\theta_2})$ and $\psi \circ \phi$, $\phi \circ \psi$ are identity maps). (Hint: Use the standard trigonometric identities to show that $e^{i\theta_1}e^{i\theta_2} = e^{i(\theta_1+\theta_2)}$.)

(5) Show that if $A \in o(n)$ ($A \in u(n)$), then the diagonal elements of A are zero (purely imaginary).

(6) Show how we can think of $M_n(\mathbf{C})$ as a $(2n)^2$-dimensional real vector space. (Recall that we already did this for $n = 1$.) Now show that $u(n)$ is a subspace.

(7) Show that $su(n)$ is a Lie Algebra and that its dimension as a (real) vector space is $n^2 - 1$.

(8) The *special linear group* is the set

$$SL(n, \mathbf{R}) = \{A \in M_n(\mathbf{R}) \mid \det A = 1\}$$

whose group operation is matrix multiplication. Show that this is, indeed, a group and find a matrix $A \in SL(n, \mathbf{R}) - O(n)$.

(9) Show that

$$sl(n, \mathbf{R}) = \{A \in M_n(\mathbf{R}) \mid \text{trace } A = 0\}$$

is a Lie Algebra. (This is the Lie Algebra associated to the Lie Group $SL(n, \mathbf{R})$.)

Curves in Matrix Groups.

A *curve* in the real normed vector space V is a continuous map $\gamma : (a, b) \to V$. For $c \in (a, b)$, we say γ is *differentiable* at c if $\lim_{h \to 0} \dfrac{\gamma(c+h) - \gamma(c)}{h}$ exists, and we denote this limit by $\gamma'(c)$.

Given the subset $W \subseteq V$, we say that γ is a curve in W if $\gamma(t) \in W$ for every $t \in (a, b)$. If G is a multiplicative group in V (e.g., $V = M_n(\mathbf{R})$ or $M_n(\mathbf{C})$, and $G = O(n)$ or $U(n)$, and if γ, σ are two curves in G, then we define their *product curve* in G by

$$(\gamma\sigma)(t) = \gamma(t)\sigma(t)\ .$$

(10) If γ and σ are both differentiable at $c \in (a, b)$, show that $\gamma\sigma$ is also differentiable at c, and

$$(\gamma\sigma)'(c) = \gamma(c)\sigma'(c) + \gamma'(c)\sigma(c)\ .$$

(11) Let β be a differentiable curve $\beta : (-a, a) \to O(n)$ with $\beta(0) = I$. Show that $\beta'(0)$ is skew-symmetric. For $\beta : (-a, a) \to U(n)$ with $\beta(0) = I$, show that $\beta'(0)$ is skew-Hermitian.

(Exercise (11) should provide a hint as to how we associate a Lie Algebra to a Lie Group.)

E. Stable Subspaces for Unitary and Orthogonal Groups

We begin with two easy observations about the unitary matrix $A \in U(n)$.

(1) *If the subspace $W \subseteq \mathbf{C}^n$ is A-stable, then $A|_W$ is unitary.*

(2) *If* $W \subseteq \mathbf{C}^n$ *is A-stable, then the orthogonal complement of W*

$$W^\perp = \{z \in \mathbf{C}^n \mid \langle z, y\rangle = 0 \quad \text{for all} \quad y \in W\}$$

is also A-stable.

Proof: Note that $A|_W \in \mathrm{End}(W)$ since W is A-stable. Also

$$\langle xA, yA\rangle = \langle x, y\rangle$$

for all $x, y \in W$ so that $A|_W$ is unitary on W.

Now choose $z \in W^\perp$ and note that for any $y \in W$ we have

$$\langle zA, y\rangle = \langle z, y\,{}^t\bar{A}\rangle = \langle z, yA^{-1}\rangle \,.$$

But A is an isomorphism of W onto W so $yA^{-1} \in W$. Thus $\langle zA, y\rangle = 0$ for all $y \in W$, showing that $zA \in W^\perp$. ∎

Proposition 12 *Given* $A \in U(n)$, *there exists an orthonormal basis for* $\mathbf{C}^n$ *consisting of eigenvectors for A.*

Proof: We show this by induction on n. For $n = 1$, the result is trivial; so suppose $n > 1$. Since the characteristic polynomial of A factors completely, A has at least one eigenvalue and hence an eigenvector v_1. Then

$$u_1 = \frac{1}{\|v_1\|} v_1$$

is a unit vector.

Let $W = \mathrm{Span}(u_1)$, so that $\mathbf{C}^n = W \oplus W^\perp$ by exercise (1) of Section IV B. Note that $\dim W^\perp = n - 1$ (since $\dim W = 1$), and $A|_{W^\perp}$ is unitary on $W^\perp$ (by our observations). By induction, $W^\perp$ has an orthonormal basis $\{u_2, \ldots, u_n\}$ of eigenvectors for $A|_{W^\perp}$.

Thus $\{u_1, \ldots, u_n\}$ is the desired basis. ∎

Corollary $A \in U(n)$ is diagonalizable.

Exercise. Given $A \in U(n)$, show that there is a $B \in U(n)$ such that BAB^{-1} is diagonal.

It is tempting to assert that Proposition 12 should also hold for $A \in O(n)$ since the observations (1) and (2) still hold for an orthogonal matrix. But we have seen that $A \in O(n)$ need not have an eigenvalue (in $\mathbf{R}$). For example,

$$A = \begin{pmatrix} 0 & 1 \\ -1 & 0 \end{pmatrix} \in O(2)$$

is not diagonalizable.

However, if $A \in M_n(\mathbf{R})$ is symmetric, then our observations still hold.

Exercise. State and prove appropriate versions of observations (1) and (2) for a real symmetric matrix.

So if we can show that a symmetric matrix always has an eigenvector, then the proof of Proposition 12 can easily be adjusted to prove the following.

Proposition 13 *If $A \in M_n(\mathbf{R})$ is symmetric, then there is an orthonormal basis for $\mathbf{R}^n$ consisting of eigenvectors for A.*

We show that a symmetric matrix has an eigenvector in the next two propositions.

Proposition 14 *Let $A \in M_n(\mathbf{R})$ be symmetric and suppose $\lambda \in \mathbf{C}$ is an eigenvalue of A. (Thinking of A as an element of $M_n(\mathbf{C})$, we know such a λ exists.) Then $\lambda \in \mathbf{R}$.*

Proof: Let $x \in \mathbf{C}^n$ be an eigenvector for A corresponding to λ. Using the standard inner product on $\mathbf{C}^n$, we see that

$$\lambda\langle x, x\rangle = \langle \lambda x, x\rangle = \langle xA, x\rangle = \langle x, x\,{}^t\bar{A}\rangle\,.$$

But A is real and symmetric, so ${}^t\bar{A} = A$. Thus

$$\langle x, x\,{}^t\bar{A}\rangle = \langle x, xA\rangle = \langle x, \lambda x\rangle = \bar{\lambda}\langle x, x\rangle$$

and we see that

$$\lambda\langle x, x\rangle = \bar{\lambda}\langle x, x\rangle\,.$$

Since x is an eigenvector, it is nonzero and so $\langle x, x\rangle \neq 0$. Hence

$$\lambda = \bar{\lambda}$$

implying that λ is real. ∎

Even though this proposition shows that the eigenvalues of A are all real, we cannot immediately conclude that the corresponding eigenvectors are in $\mathbf{R}^n$ (because we had to work in $\mathbf{C}^n$).

Proposition 15 *The symmetric matrix A has an eigenvector in $\mathbf{R}^n$.*

Proof: Let $\lambda \in \mathbf{R}$ be an eigenvalue of A and let $z \in \mathbf{C}^n$ be a corresponding eigenvector. We can write z uniquely as

$$z = x + iy$$

where $x, y \in \mathbf{R}^n$. (Exercise. Show this.) And so

$$\lambda x + i\lambda y = \lambda z = zA = xA + iyA\,.$$

Since λ is real and since the real and imaginary parts of a vector are unique, we must have

$$xA = \lambda x \quad \text{and} \quad yA = \lambda y \,.$$

But z is nonzero, so one of x or y must be nonzero. ∎

Exercise. Given the symmetric matrix $A \in M_n(\mathbf{R})$, show that there is a $B \in O(n)$ such that BAB^{-1} is diagonal.

Now we return to orthogonal matrices and show that even though

$$A = \begin{pmatrix} \cos\theta & -\sin\theta \\ \sin\theta & \cos\theta \end{pmatrix} \in SO(2)$$

is not diagonalizable (unless $\theta = k\pi$ for $k \in \mathbf{Z}$), this is about the worst case. That is

Proposition 16 *Given $A \in O(n)$, $\mathbf{R}^n$ contains A-stable subspaces of dimensions one and/or two such that $\mathbf{R}^n$ is the internal direct sum of these subspaces.*

Proof: Again, we use induction on n. For $n = 1$, there is nothing to prove; so assume $n > 1$.

Lemma *There exists an A-stable subspace $W \subseteq \mathbf{R}^n$ with* $\dim W \in \{1, 2\}$.

Proof: $B = A + {}^tA$ is symmetric, so B has an eigenvector $x \in \mathbf{R}^n$ (by Proposition 15). Let $\lambda \in \mathbf{R}$ be the corresponding eigenvalue, and consider the two vectors x, xA.

If x and xA are linearly dependent, then $xA = rx$ for some $r \in \mathbf{R}$. Thus

$$W = \{tx \mid t \in \mathbf{R}\}$$

is A-stable and $\dim W = 1$.

If x and xA are linearly independent, let

$$W = \mathrm{Span}(x, xA) \,.$$

Note that

$$\lambda x = xB = xA + x\,{}^tA = xA + xA^{-1}$$

and so

$$xA^2 = \lambda xA - x \,.$$

Now choose $w \in W$ and write this as $w = ax + bxA$ for some $a, b \in \mathbf{R}$. But then

$$\begin{aligned} wA &= (ax + bxA)A = axA + bxA^2 \\ &= axA + b(\lambda xA - x) = (-b)x + (a + b\lambda)xA \in W \end{aligned}$$

so that W is A-stable.

This proves the lemma. ∎

Returning to the proof of Proposition 16, we let W be an A-stable subspace as in the lemma, and note that $\mathbf{R}^n = W \oplus W^\perp$.

By our observations, $A|_{W^\perp}$ is orthogonal on $W^\perp$ (which is of dimension $< n$); so by induction, $W^\perp$ can be written as the direct sum of A-stable subspaces of dimensions one and/or two. ∎

Given $A \in O(n)$, Proposition 16 allows us to restrict our attention to the A-stable subspace $W \subseteq \mathbf{R}^n$ which is of dimension one or two. So by Proposition 13 of Section III E, we only have to look at 1×1 and 2×2 orthogonal matrices.

$A \in O(1)$ must be (1) or (-1) since $|\det A| = 1$. What about $A \in O(2)$?

Exercise. Show that $A \in O(2)$ can be written as

$$\begin{pmatrix} \cos\theta & -\sin\theta \\ \sin\theta & \cos\theta \end{pmatrix} \text{ or } \begin{pmatrix} -\cos\theta & \sin\theta \\ \sin\theta & \cos\theta \end{pmatrix}$$

for some $\theta \in \mathbf{R}$.

Note that $A = \begin{pmatrix} -\cos\theta & \sin\theta \\ \sin\theta & \cos\theta \end{pmatrix}$ is diagonalizable since its characteristic polynomial is $x^2 - 1$. (Exercise. Show this.) So A is similar to $\begin{pmatrix} 1 & 0 \\ 0 & -1 \end{pmatrix}$.

Combining Proposition 16 with these observations gives the following.

Corollary Given $A \in O(n)$, there is a nonsingular matrix C such that CAC^{-1} is a block diagonal matrix of the form

$$\begin{pmatrix} M_1 & & & & 0 \\ & M_2 & & & \\ & & M_3 & & \\ & & & \ddots & \\ 0 & & & & M_k \end{pmatrix} \tag{22}$$

where $M_1 = I, M_2 = -I, M_3 = \begin{pmatrix} \cos\theta_3 & -\sin\theta_3 \\ \sin\theta_3 & \cos\theta_3 \end{pmatrix}, \ldots,$
$M_k = \begin{pmatrix} \cos\theta_k & -\sin\theta_k \\ \sin\theta_k & \cos\theta_k \end{pmatrix}$.

Exercises

(1) Show that $\begin{pmatrix} \cos\theta & -\sin\theta \\ \sin\theta & \cos\theta \end{pmatrix}$ is not diagonalizable unless $\theta = k\pi$ for $k \in \mathbf{Z}$.

(2) Show that

$$A = \begin{pmatrix} \frac{2}{3} & \frac{1}{3} & \frac{2}{3} \\ \frac{1}{\sqrt{2}} & 0 & \frac{-1}{\sqrt{2}} \\ \frac{-1}{3\sqrt{2}} & \frac{4}{3\sqrt{2}} & \frac{-1}{3\sqrt{2}} \end{pmatrix}$$

is orthogonal and find a matrix of the form (22) which is similar to A.

(3) $A \in M_n(\mathbf{C})$ is said to be a *normal matrix* if $A\,{}^t\bar{A} = {}^t\bar{A}A$ (or $A \in M_n(\mathbf{R})$ is normal if $A{}^tA = {}^tAA$). Note that Hermitian and unitary matrices are special cases of normal matrices. Show that

$$A = \begin{pmatrix} 1 & i \\ 1 & 3+2i \end{pmatrix}$$

is normal.

(4) Suppose $A \in M_n(\mathbf{C})$ is normal. Show that

(i) A is Hermitian $\Leftrightarrow$ the eigenvalues of A are real, and

(ii) A is unitary $\Leftrightarrow$ the eigenvlaues of A are all of absolute value 1.

(5) Given $A \in M_n(\mathbf{C})$ and an A-stable subspace $W \subseteq \mathbf{C}^n$, show that $W^\perp$ is ${}^t\bar{A}$-stable.

(6) Let $A \in M_n(\mathbf{C})$ be normal. Show that

(i) $xA = 0 \Leftrightarrow x\,{}^t\bar{A} = 0$,

(ii) $A - \lambda I$ is normal for any $\lambda \in \mathbf{C}$, and

(iii) if $xA = \lambda x$, then $x\,{}^t\bar{A} = \bar{\lambda}x$.

(7) If $A \in M_n(\mathbf{C})$ is normal, show that there is an orthonormal basis for $\mathbf{C}^n$ consisting of eigenvectors for A. (Hint: Use exercises (5) and (6) to prove appropriate versions of the observations (1) and (2).)

(8) Show that any endomorphism of a finite-dimensional complex inner product space can be represented by a triangular matrix.

(9) Suppose the $n \times n$ matrix A is normal. If x and y are eigenvectors belonging to distinct eigenvalues of A, show that x and y are orthogonal.

(10) Recall that $GL(n, \mathbf{R}) = \{A \in M_n(\mathbf{R}) \mid \det A \neq 0\}$. If $P(n)$ denotes the set of $n \times n$ positive definite matrices (see exercise (8) of Section IV C), show that

$$GL(n, \mathbf{R}) = O(n) \times P(n) .$$

That is, given $A \in GL(n, \mathbf{R})$, there exist $B \in O(n)$ and $C \in P(n)$ such that

$$A = BC .$$

Furthermore, B and C are unique. (Hint: $N = A\,{}^tA$ is symmetric, so there is an orthogonal matrix Q such that QNQ^{-1} is diagonal; but the diagonal elements are positive since N is positive definite. Thus there is a (real) diagonal matrix S such that $S^2 = QNQ^{-1}$. Let $C = Q^{-1}SQ$.)

(11) Show that $GL(n, \mathbf{C}) = U(n) \times P(n)$.

Chapter V

Normed Algebras

A. The Normed Algebras R and C

We have seen that normed vector spaces are the same as inner product spaces precisely when the norm satisfies the parallelogram law.

We now begin our search for normed algebras. Let A be a finite-dimensional real vector space and choose a basis $\{v_1, \ldots, v_n\}$ for A. Suppose that A has a multiplication which makes it an algebra (which is not necessarily commutative or associative) with unit element $v_1 = 1$.

If we define an inner product on A by $\langle v_i, v_j \rangle = \delta_{ij}$, then we get the associated norm on A as usual; i.e.

$$\|v\| = \sqrt{\langle v, v \rangle} .$$

We say that A is a *normed algebra* if

$$\|uv\| = \|u\| \, \|v\| \quad \text{for all } u, v \in A . \tag{1}$$

Note that unlike the case of a normed vector space, we require the norm on an algebra to come from an inner product.

We begin with the observation that $\mathbf{R}$ is a normed algebra. Its norm is $\|x\| = \sqrt{x^2}$ and we easily check that we do have $\|xy\| = \|x\| \, \|y\|$. So $\mathbf{R}$ is our first example.

Next we claim that the vector space $\mathbf{R}^2$ with its usual norm $\|x\| = \sqrt{x_1^2 + x_2^2}$ can be made into a normed algebra by an appropriate definition of multiplication. Indeed, the multiplication will be unique in the following sense. We consider

$$\mathbf{R} = \{(x, 0)\} \subseteq \mathbf{R}^2$$

and we insist that our multiplication on $\mathbf{R}^2$ extend that on $\mathbf{R}$; i.e., $(a, 0)(b, 0) = (ab, 0)$, and $(1, 0)$ is the unit element for the multiplication on $\mathbf{R}^2$. Then there is only one such multiplication making $\mathbf{R}^2$ into a normed algebra.

To see this, we take the standard basis $\{(1,0),(0,1)\}$ for $\mathbf{R}^2$ and denote these vectors by

$$1=(1,0) \text{ and } i=(0,1).$$

Then any $(a,b) \in \mathbf{R}^2$ can be uniquely written as $a1+bi$ or, since 1 is the multiplicative identity, as $a+bi$. We always want multiplication to distribute over addition so we must have

$$(2) \quad (a+bi)(c+di) = ac+bci+adi+bdi^2 = ac+(bc+ad)i+bdi^2$$

so we just need to define i^2 so that (1) is satisfied. We have $\|i\| = \sqrt{1^2+0^2} = 1$ so that $\|i\|\,\|i\| = \|i^2\| = 1$ by (1). This shows we can write $i^2 = u+wi$ with $u,w \in \mathbf{R}$ and $u^2+w^2=1$. Then

$$\begin{aligned} 1=\|i^3\|^2 &= \|(u+wi)i\|^2 = \|ui+w(u+wi)\|^2 \\ &= \|uw+(u+w^2)i\|^2 = u^2w^2+u^2+2uw^2+w^4 , \end{aligned}$$

and putting in $u^2 = 1-w^2$ gives

$$\begin{aligned} 1 &= (1-w^2)w^2+(1-w^2)+2uw^2+w^4 \\ &= w^2-w^4+1-w^2+2uw^2+w^4 \\ &= 1+2uw^2 . \end{aligned}$$

Thus we must have $2uw^2=0$ so that $u=0$ or $w=0$.

If $u=0$, then $w=1$ or $w=-1$. If $w=1$ (and $u=0$), then $i^2=i$ which implies $i=1$ (see Exercise (3)); i.e., $(1,0)=(0,1)$, which is false. If $u=0$ and $w=-1$, we get $i^2=-i$, implying that $i=-1$, also a contradiction. Thus we cannot have $u=0$.

So $w=0$ and $i^2=1$ or $i^2=-1$. If $i^2=1$, we get

$$0=\|i^2-1\| = \|i+1\|\,\|i-1\| = \sqrt{2}\sqrt{2} = 2 ,$$

a contradiction. Thus $i^2=-1$ and (2) becomes

$$(2') \qquad (a+bi)(c+di) = (ac-bd)+(ad+bc)\,i.$$

<u>Exercise</u>. Show that this multiplication is associative, commutative, extends that of $\mathbf{R} = \{a+0i\} \subseteq \mathbf{R}^2$, and preserves norms. (Hint: Recall the exercises of Chapter 0.)

The previous example is typical of the situation in which we are interested. That is, if A is a normed algebra, can we extend A to a larger normed algebra B such that the multiplication of B extends that of A and the norm on B restricts to that on A? If such a B exists, we say that it has A as a *subalgebra*.

An application of (2′) to number theory is

Theorem *Let $S \subseteq \mathbf{Z}$ be those integers which may be written as a sum of two squares (e.g., $29 \in S$ since $29 = (2)^2 + (5)^2$). Then S is multiplicatively closed.*

Proof: Suppose $\rho = a^2 + b^2$ and $\sigma = c^2 + d^2$ are in S.

We need to show $\rho\sigma \in S$. Set $\alpha = a + bi$, $\beta = c + di$. Then we have

$$\begin{aligned} \rho\sigma &= (a^2+b^2)(c^2+d^2) = \|\alpha\|^2\|\beta\|^2 = \|\alpha\beta\|^2 \\ &= (ac-bd)^2 + (ad+bc)^2 . \quad \blacksquare \end{aligned}$$

Of course, the normed algebra we found on $\mathbf{R}^2$ is just the field $\mathbf{C}$ of complex numbers. So now we ask if we can extend this normed algebra on $\mathbf{C}$ to a normed algebra on $\mathbf{R}^3$. The answer is a resounding NO. In fact, we cannot even extend the multiplication on $\mathbf{C}$ to an associative multiplication on $\mathbf{R}^3$. (Later we will get this result without assuming associativity.)

Suppose we can extend the multiplication on $\mathbf{C}$ to $\mathbf{R}^3$. We extend the basis $\{1, i\}$ of $\mathbf{C}$ to the basis

$$\{1 = (1,0,0), \quad i = (0,1,0), \quad j = (0,0,1)\} \quad \text{for } \mathbf{R}^3$$

and note that we must have

$$ij = a + bi + cj \quad \text{with } a, b, c \in \mathbf{R}.$$

If we assume that the extension of $\mathbf{C}$ to $\mathbf{R}^3$ is an associative algebra, then

$$-j = i^2 j = i(ij) = ai - b + c(a + bi + cj) = (ac - b) + (a + bc)i + c^2 j$$

and equating coefficients of j gives $-1 = c^2$, contradicting $c \in \mathbf{R}$.

Next we turn our attention to $\mathbf{R}^4$. We shall see that we can make $\mathbf{R}^4$ into a normed algebra which is associative but not commutative. Our proof of this will be constructive; that is, we will use some properties of a general normed algebra to decide how we should define a multiplication on $\mathbf{R}^4$.

We use the basis

$$\{1 = (1,0,0,0),\ i = (0,1,0,0),\ j = (0,0,1,0),\ k = (0,0,0,1)\}$$

along with the standard norm (i.e., if $x = x_1 + x_2 i + x_3 j + x_4 k$ then $\|x\| = \sqrt{x_1^2 + x_2^2 + x_3^2 + x_4^2}$). Also, we want $\mathbf{C}$ to be a subalgebra, so we insist that $i^2 = -1$.

Let $\mathbf{R} = \text{Span}(1)$ so that its orthogonal complement is

$$\mathbf{R}^\perp = \text{Span}(i, j, k) .$$

Then any $q \in \mathbf{R}^4$ can be uniquely written as $q = a + \beta$ with $a \in \mathbf{R}$ and $\beta \in \mathbf{R}^\perp$. We call $a = \text{Re}(q)$ the *real part* of q and $\beta = \text{Im}(q)$ the *imaginary part* of q.

We define the *conjugate* of $q = a + \beta$ to be

$$\bar{q} = a - \beta .$$

Then clearly

$$\bar{\bar{q}} = q,\ \bar{\beta} = -\beta,\ \mathrm{Re}(q) = \frac{q + \bar{q}}{2}, \quad \text{and } \mathrm{Im}(q) = \frac{q - \bar{q}}{2} .$$

Also, for $q_1, q_2 \in \mathbf{R}^4$ we have

$$\overline{q_1 + q_2} = \bar{q}_1 + \bar{q}_2 .$$

Exercise. Show that these conjugation formulas hold.

We want to prove that for any $q \in \mathbf{R}^4$, we have

$$q\bar{q} = \|q\|^2 . \tag{3}$$

It turns out to be fairly easy to prove (3) for any normed algebra, so we will do this in the next section.

After proving (3) in the general setting, we will finish making $\mathbf{R}^4$ into a normed algebra. We call this normed algebra the *quaternions* and denote it by $\mathbf{H}$. (The $\mathbf{H}$ is in honor of W.R. Hamilton, who invented the quaternions in 1843.)

But first, we observe that if $\beta \in \mathbf{R}^\perp$ is any unit vector, then we have

$$\beta\bar{\beta} = \|\beta\|^2 = 1$$

by (3). But $\beta\bar{\beta} = \beta(-\beta) = -\beta^2$ so that

$$\beta^2 = -1 . \tag{4}$$

In particular, we must have $j^2 = k^2 = -1$.

Next, we note that ij must be in $\mathbf{R}^\perp$ (we will see in Proposition 3 that $\langle xw, y\rangle = \langle x, y\bar{w}\rangle$ for any $x, y, w \in \mathbf{R}^4$, so $\langle ij, 1\rangle = \langle i, \bar{j}\rangle = -\langle i, j\rangle = 0$; so by (4), we also have

$$ij\,ij = (ij)^2 = -1 \tag{5}$$

since ij is a unit vector. (At this stage, we are assuming that we can construct an associative multiplication.) Multiplying (5) on the left by i and on the right by j gives

$$ji = -ij .$$

Similarly, we see that $jk = -kj$ and $ki = -ik$.

Consequently, $\mathbf{H}$ is not commutative.

Exercises

(1) Show that the unit element of any normed algebra is a unit vector.

(2) Suppose that A is a normed algebra and that $a \in A$ is a unit (i.e., there is a (unique) $a^{-1} \in A$ such that $aa^{-1} = a^{-1}a = 1$). Show that

$$\|a^{-1}\| = \|a\|^{-1} .$$

(3) Suppose a and b are elements of a normed algebra. Show that if $ab = 0$, then either $a = 0$ or $b = 0$. Use this to show that if $i^2 = i$, then $i = 1$ (hence $i^2 \neq i$).

(4) What is wrong with the following "proof" that j and k are linearly dependent? Since $j^2 = k^2 = -1$, we have $j^2 - k^2 = 0$. But $x^2 - y^2$ factors as $(x+y)(x-y)$, so we get $(j+k)(j-k) = 0$. By exercise (3), either $j = -k$ or $j = k$.

(5) Read "Hamilton's Discovery of Quaternions" by B.L. van der Waerden in Mathematics Magazine, vol. 49, #5 (1976), pp. 227-234.

B. Some General Results, Quaternions

Let A be a normed algebra with basis $\{v_1, \ldots, v_n\}$ such that $v_1 = 1$ and the inner product on A is

$$\langle x, y \rangle = x_1 y_1 + \cdots + x_n y_n$$

where $x = x_1 v_1 + \cdots + x_n v_n$, $y = y_1 v_1 + \cdots + y_n v_n$.

Recall that the *commutator* of x and y is

$$[x, y] = xy - yx,$$

and the *associator* of x, y, z is

$$[x, y, z] = (xy)z - x(yz).$$

As before, $\mathbf{R} = \text{Span}(1)$ and its orthogonal complement is $\mathbf{R}^{\perp} = \text{Span}(v_2, \ldots, v_n)$. Also, we can write $x \in A$ uniquely as $x = a + \beta$ with *real part* $a = \text{Re}(x) \in \mathbf{R}$ and *imaginary part* $\beta = \text{Im}(x) \in \mathbf{R}^{\perp}$. We define the *conjugate* of x to be

$$\bar{x} = a - \beta ,$$

and it is easy to show that

$$\bar{\bar{x}} = x,\ \bar{\beta} = -\beta,\ a = \frac{x+\bar{x}}{2},\ \beta = \frac{x-\bar{x}}{2},\ \text{and}\ \overline{x+y} = \bar{x}+\bar{y}\ .$$

Notice that if $x \in \mathbf{R}$ ($\Leftrightarrow \mathrm{Im}(x) = 0$), then we can consider x to be a scalar. In fact, A need not be normed for this to hold.
Exercise. Let B be any real algebra with unit element. Show that $a \in \mathrm{Span}(1)$ can be considered to be a scalar.

Proposition 1 *Suppose B is a real algebra with unit element (not necessarily normed).*

(i) *For $x, y \in B$, if one of them is real, then $[x,y] = 0$.*

(ii) *For $x, y, z \in B$, if any one of them is real, then $[x,y,z] = 0$.*

Proof: (i) and (ii) both follow from the preceding exercise and the definition of an algebra. ∎

From now on, our algebras will be normed unless we state otherwise.

Proposition 2 *For any $x, y, w \in A$, we have*

(i) $\langle xw, yw\rangle = \langle x, y\rangle\|w\|^2 \quad (= \langle x,y\rangle\langle w,w\rangle)$, *and*

(ii) $\langle wx, wy\rangle = \langle x, y\rangle\|w\|^2$.

Proof: We prove (i); (ii) can be proved similarly. First we obtain an identity in x and w. We have $\|xw\|^2 = \|xw\|\,\|xw\| = \|x\|\,\|w\|\,\|x\|\,\|w\| = \|x\|^2\|w\|^2$ so that

$$\langle xw, xw\rangle = \langle x, x\rangle\langle w, w\rangle\ . \tag{6}$$

Now we prove (i) by polarizing the identity (6). We put $x+y$ in (6) in place of x to get

$$\langle (x+y)w, (x+y)w\rangle = \langle x+y, x+y\rangle\langle w,w\rangle\ .$$

We expand both inner products (by bilinearity and symmetry) to get

$$\langle xw, xw\rangle + 2\langle xw, yw\rangle + \langle yw, yw\rangle = (\langle x,x\rangle + 2\langle x,y\rangle + \langle y,y\rangle)\langle w,w\rangle\ .$$

Using (6) to cancel some terms in this last equation gives

$$2\langle xw, yw\rangle = 2\langle x,y\rangle\langle w,w\rangle\ ,\ \text{proving (i)}.$$

Exercise. Prove (ii). ∎

Proposition 3 *For any $x, y, w \in A$, we have*

(i) $\langle xw, y\rangle = \langle x, y\bar{w}\rangle$, *and*

(ii) $\langle wx, y\rangle = \langle x, \bar{w}y\rangle$.

Proof: Again, we prove (i).

Let $w = c + \delta$ with $c \in \mathbf{R}$ and $\delta \in \mathbf{R}^\perp$. Then

$$\begin{aligned} \langle xw, y\rangle &= \langle cx, y\rangle + \langle x\delta, y\rangle, \quad \text{and} \\ \langle x, y\bar{w}\rangle &= \langle x, cy\rangle + \langle x, y\bar{\delta}\rangle. \end{aligned}$$

Since $\langle cx, y\rangle = \langle x, cy\rangle$, it suffices to prove (i) in the case that $w \in \mathbf{R}^\perp$ (so that $\bar{w} = -w$). With the assumption that $w \in \mathbf{R}^\perp$ we claim

$$\langle x, y\rangle(1 + \|w\|^2) = \langle x(1+w), y(1+w)\rangle\,. \tag{7}$$

By Proposition 2, the right hand side of (7) equals

$$\langle x, y\rangle\langle 1+w, 1+w\rangle = \langle x, y\rangle(\langle 1, 1\rangle + \langle 1, w\rangle + \langle w, 1\rangle + \langle w, w\rangle)\,.$$

But $\langle 1, w\rangle = \langle w, 1\rangle = 0$, so this equals $\langle x, y\rangle(1 + \|w\|^2)$ as claimed.

By proposition 2, we can rewrite (7) as

$$\langle x, y\rangle + \langle xw, yw\rangle = \langle x, y\rangle + \langle xw, y\rangle + \langle x, yw\rangle + \langle xw, yw\rangle\,.$$

It follows that $\langle xw, y\rangle = -\langle x, yw\rangle = \langle x, y(-w)\rangle = \langle x, y\bar{w}\rangle$, proving (i).

Exercise. Prove (ii). ∎

Next we establish an important property of conjugation.

Proposition 4 *For $x, y \in A$, we have*

$$\overline{xy} = \bar{y}\bar{x}.$$

Proof: For any $x, y, z, w \in A$ we use (i) and (ii) of Proposition 3 repeatedly to get

$$\begin{aligned} \langle \bar{z}\bar{w}, \overline{xy}\rangle &= \langle \bar{z}, \overline{xy}w\rangle = \langle (xy)\bar{z}, w\rangle = \langle xy, wz\rangle = \langle y, \bar{x}(wz)\rangle \\ &= \langle y(\overline{wz}), \bar{x}\rangle = \langle \overline{wz}, \bar{y}\bar{x}\rangle. \end{aligned}$$

Now take $w = 1$ to get $\langle \bar{z}, \overline{xy}\rangle = \langle \bar{z}, \bar{y}\bar{x}\rangle$. Since $\langle\,,\rangle$ is nondegenerate, this proves Proposition 4. ∎

We are finally ready to prove formula (3) (i.e., $x\bar{x} = \|x\|^2$ for any $x \in A$).

Proposition 5 *For $x, y \in A$, we have $\langle x, y\rangle = \mathrm{Re}(x\bar{y}) = \frac{1}{2}(x\bar{y} + y\bar{x})$ and so*

$$x\bar{x} = \|x\|^2\,.$$

Proof: Recall that $\{v_1, \ldots, v_n\}$ is our basis of A, and we write $x = x_1 v_1 + \cdots + x_n v_n$, $y = y_1 v_1 + \cdots + y_n v_n$.

First we consider $v_i v_j$ when $i \neq j$. If $j = 1$, then $v_i v_j = v_i \in \mathbf{R}^\perp$; so assume $j > 1$. By Proposition 3,

$$\langle v_i v_j, 1\rangle = \langle v_i, \bar{v}_j\rangle = \langle v_i, -v_j\rangle = -\langle v_i, v_j\rangle = 0 .$$

Thus $v_i v_j \in \mathbf{R}^\perp$ whenever $i \neq j$.

Next we note that

$$\|\bar{v}_i\|^2 = \langle \bar{v}_i, \bar{v}_i\rangle = \langle 1, \bar{v}_i v_i\rangle = \langle v_i, v_i\rangle = 1$$

by Proposition 3, so $\bar{v}_i$ is a unit vector. But then $v_i\bar{v}_i$ is also a unit vector; furthermore, $v_i\bar{v}_i$ is real since

$$\overline{w\bar{w}} = \bar{\bar{w}}\bar{w} = w\bar{w} \quad \text{for any} \quad w \in A$$

by Proposition 4. Thus $v_i\bar{v}_i \in \{1, -1\}$ for all i.

For $i > 1$, we have $\bar{v}_i = -v_i$ so that $-v_i^2 = v_i\bar{v}_i \in \{1, -1\}$. If $v_i^2 = 1$, then

$$0 = v_i^2 - 1 = (v_i + 1)(v_i - 1)$$

which contradicts the fact that neither $v_i + 1$ nor $v_i - 1$ is zero. Hence $v_i^2 = -1$ if $i > 1$.

Now we use these observations to calculate

$$\begin{aligned} x\bar{y} &= (x_1 + x_2 v_2 + \cdots + x_n v_n)(y_1 - y_2 v_2 - \cdots - y_n v_n) \\ &= x_1 y_1 + x_2 y_2 + \cdots + x_n y_n + y_1(x_2 v_2 + \cdots + x_n v_n) - \sum_{i \neq j \neq 1} x_i y_j v_i v_j \end{aligned}$$

so that $\mathrm{Re}(x\bar{y}) = x_1 y_1 + \cdots + x_n y_n$. But this is precisely $\langle x, y\rangle$. The fact that $\mathrm{Re}(x\bar{y}) = \frac{1}{2}(x\bar{y} + y\bar{x})$ follows from the fact that $\mathrm{Re}(w) = \frac{1}{2}(w + \bar{w})$ and Proposition 4.

Finally, we see that

$$\|x\|^2 = \langle x, x\rangle = \mathrm{Re}(x\bar{x}) = x\bar{x} . \quad \blacksquare$$

Corollary *If $\beta \in \mathbf{R}^\perp$ is a unit vector, then $\beta^2 = -1$. Consequently, when A is associative, $v_i v_j = -v_j v_i$ for distinct $v_i, v_j \in \{v_2, \ldots, v_n\}$.*
Proof: Exercise. $\blacksquare$

Quaternions

We have the basis $\{1, i, j, k\}$ for $\mathbf{R}^4$ and we know that

$$\begin{gathered} i^2 = j^2 = k^2 = -1 , \\ ji = -ij, \quad jk = -kj, \quad \text{and} \quad ki = -ik \end{gathered}$$

(assuming that our multiplication will be associative). But we still need to define ij, jk, and ki.

Choose $w \in \mathbf{R}^4$ and consider the map $L_w : \mathbf{R}^4 \to \mathbf{R}^4$ defined by

$$L_w(x) = wx \,.$$

L_w is called *left multiplication* by w. This is a linear map since

$$L_w(ax + by) = w(ax + by) = awx + bwy = aL_w(x) + bL_w(y) \,.$$

That is, $L_w \in \mathrm{End}(\mathbf{R}^4)$. Letting $w = i$, we see that $L_{(-i)}$ is the inverse of L_i since

$$L_{(-i)} \circ L_i(x) = L_{(-i)}(ix) = (-i)(ix) = x$$

and

$$L_i \circ L_{(-i)}(x) = x \,;$$

so L_i is an isomorphism of $\mathbf{R}^4$. By Proposition 2, we have

$$\langle L_i(x), L_i(y)\rangle = \langle x, y\rangle \|i\|^2 = \langle x, y\rangle$$

so that L_i is an orthogonal map (i.e., L_i takes $\{1, i, j, k\}$ to another orthonormal basis of $\mathbf{R}^4$). Now L_i sends 1 to i and it sends i to -1. Thus L_i must take $\{j, k\}$ to an orthonormal basis for $\mathrm{Span}(j, k)$. That is,

$$L_i(j) = ij = aj + bk \quad \text{with } a, b \in \mathbf{R}$$

such that $a^2 + b^2 = 1$ and $b \neq 0$. (If $b = 0$, then we have $ij \in \{j, -j\}$ which implies that i is real.) Similarly,

$$L_i(k) = ik = cj + dk \quad \text{with } c, d \in \mathbf{R} \,,$$

$c^2 + d^2 = 1$, and $c \neq 0$. Then we have

$$\begin{aligned} iL_i(j) = -j &= aij + bik \\ &= a(aj + bk) + b(cj + dk) \\ &= (a^2 + bc)j + (ab + bd)k \,. \end{aligned}$$

Equating coefficients of j and k yields

$$a^2 + bc = -1 \tag{8}$$

and

$$b(a + d) = 0 \,. \tag{9}$$

Since $b \neq 0$, (9) becomes

$$a = -d \,.$$

Now L_i is orthogonal, so we have

$$0 = \langle L_i(j), L_i(k)\rangle = \langle aj + bk, cj + dk\rangle$$

which implies

(10) $$ac + bd = 0\,.$$

Substituting $a = -d$ into (10) gives

(11) $$a(c - b) = 0\,.$$

Now we multiply (8) by c and multiply (10) by a to obtain

(8′) $$a^2c + bc^2 = -c$$

(10′) $$a^2c + abd = 0\,.$$

Subtracting (10′) from (8′) yields

$$b(c^2 - ad) = -c$$

or, since $a = -d$,

$$b(c^2 + d^2) = -c\,.$$

But $c^2 + d^2 = 1$, so we see that

(12) $$b = -c\,.$$

From (11), we know that either $a = 0$ or $c = b$. If $c = b$, then we get $c = b = -c$ by (12); but then c must be zero, a contradiction.

Hence $0 = a = -d$ and so $b = -c \in \{1, -1\}$.

We *choose* $b = 1$ so that we have

$$ij = k \quad \text{and} \quad ik = -j\,.$$

Multiplying $ij = k$ on the right by j gives

$$kj = -i\,;$$

so we have defined $ij = k$, $jk = i$, and $ki = j$. These last three relations area easy to remember using the diagram below.

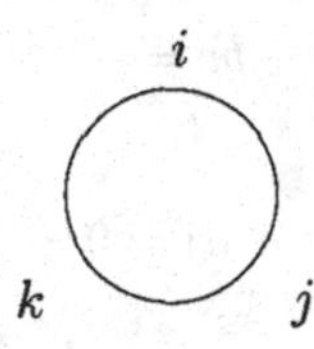

Thus we have the multiplication table

	1	i	j	k
1	1	i	j	k
i	i	-1	k	$-j$
j	j	$-k$	-1	i
k	k	j	$-i$	-1

and using this table (along with distributivity), we can multiply any two elements of $\mathbf{R}^4$.

For example, if $u = a + bi + cj + dk$ and $v = x + yi + zj + wk$, then

$$(13) \qquad \begin{aligned} uv &= (ax - by - cz - dw) + (ay + bx + cw - dz)i \\ &\quad + (az + cx + dy - bw)j + (aw + dx + bz - cy)k \,. \end{aligned}$$

As we noted before, $\mathbf{R}^4$ equipped with this multiplication is denoted by $\mathbf{H}$ and called the *quaternions.*

Exercise. Show that $\mathbf{H}$ is an associative normed algebra (which is not commutative) if its multiplication is defined by the preceding table; also, verify formula (13). Would we still get a normed algebra if we had decided to define ij as $-k$ instead of k? (Yes. In fact, this algebra is isomorphic to $\mathbf{H}$.)

So far, we have constructed the normed algebras $\mathbf{R} \subseteq \mathbf{C} \subseteq \mathbf{H}$. Can we extend $\mathbf{H}$ to a larger normed algebra? In fact we can, but we lose associativity; we can make $\mathbf{R}^8$ into a normed algebra having $\mathbf{H}$ as a subalgebra. This new normed algebra is called the *octonions* (or *Cayley numbers*) and is denoted by $\mathbf{O}$.

We will forgo a careful development of the octonians and simply define an appropriate structure on $\mathbf{R}^8$. (We remark that there are several such structures, but they give isomorphic normed algebras.)

We use the standard basis $\{1 = e_1, e_2, \ldots, e_8\}$ where $1 = (1, 0, \ldots, 0)$, etc., and the standard norm on $\mathbf{R}^8$. By the corollary after Proposition 5, we know that we must have

$$e_i^2 = -1 \quad \text{for } i \geq 2 \,;$$

also, we require that

$$e_i e_j = -e_j e_i$$

for $i, j \in \{2, \ldots, 8\}$ with $i \neq j$. (This property does not follow from the corollary. Why?) We define

$$e_2e_3 = e_4,\ e_2e_5 = e_6,\ e_2e_7 = -e_8,\ e_3e_5 = e_7, \\ e_3e_6 = e_8,\ e_4e_5 = e_8,\ e_4e_6 = -e_7$$

and we get 14 more relations by cyclically permuting each set of three subscripts (e.g., $e_3e_4 = e_2$ and $e_4e_2 = e_3$, etc.).

Thus we know how to multipy any two basis elements, so we can multiply two elements of $\mathbf{R}^8$ using distributivity.

Obviously, $\mathbf{O}$ is not commutative; to see that it is not even associative, we calculate

$$e_2e_5 \cdot e_7 = e_6e_7 = -e_4$$

and compare this to

$$e_2 \cdot e_5e_7 = e_2e_3 = e_4 .$$

<u>Exercises</u>

(1) Show that the commutator $[\,,] : A \times A \to A$ is a bilinear map on the algebra A. Now show that the associator $[\,,\,,]$ is trilinear on A.

(2) Show that if we represent $\mathbf{R}^4$ as $\mathbf{C} \oplus \mathbf{C} = \{(\alpha, \beta) \mid \alpha, \beta \in \mathbf{C}\}$ and define

$$(\alpha, \beta)(\gamma, \delta) = (\alpha\gamma - \bar{\delta}\beta, \delta\alpha + \beta\bar{\gamma}) ,$$

then this also gives $\mathbf{H}$.

(3) Now represent $\mathbf{R}^8$ as $\mathbf{H} \oplus \mathbf{H}$ and use the same definition of multiplication as in exercise (2). Show that this also gives $\mathbf{O}$.

(4) Write out a formula, analogous to (13), for the product of two octonions.

(5) Show that $\mathbf{O}$ is a normed algebra which contains $\mathbf{H}$ as a subalgebra. (Showing that $\mathbf{O}$ is normed will be easier after we have established the Cayley-Dickson formula.)

(6) Let $F \subseteq \mathbf{Z}$ be those integers which may be represented as a sum of four squares, and let $E \subseteq \mathbf{Z}$ be those integers which may be represented as a sum of eight squares (e.g., $10 = 1^2 + 1^2 + 2^2 + 2^2 \in F$). Show tht F and E are closed under multiplication.

(7) Given $q = a + bi + cj + dk \in \mathbf{H}$ where $a, b, c, d \in \mathbf{R}$, show that q and $\bar{q}$ are both roots of the real polynomial

$$p(x) = x^2 - 2ax + \|q\| .$$

(8) Consider the set of n-tuples of quaternions

$$\mathbf{H}^n = \{(q_1, \ldots, q_n) \mid q_i \in \mathbf{H}\} .$$

This is not a vector space since $\mathbf{H}$ is not a field, but we can still define the notions of scalar multiplication, addition, linear maps, etc. For example, if $a \in \mathbf{H}$ and $q = (q_1, \cdots, q_n) \in \mathbf{H}^n$, then we define

$$aq = (aq_1, \ldots, aq_n) .$$

Given a matrix of quaternions $A \in M_n(\mathbf{H})$, we define a map $\phi : \mathbf{H}^n \to \mathbf{H}^n$ by

$$\phi(q) = (q_1, \ldots, q_n)A .$$

Show that this map is linear. Also, show that the map

$$\psi(q) = A \begin{pmatrix} q_1 \\ \vdots \\ q_n \end{pmatrix}$$

is not linear since we decided to multiply by scalars on the left.

(9) Define an inner product on $\mathbf{H}^n$ by

$$\langle x, y \rangle = x_1\bar{y}_1 + \cdots + x_n\bar{y}_n$$

where $x = (x_1, \ldots, x_n)$ and $y = (y_1, \ldots, y_n)$. Show that $\langle\, , \rangle$ satisfies the following properties.

(i) $\langle x, y \rangle = \overline{\langle y, x \rangle}$

(ii) $\langle x + y, z \rangle = \langle x, z \rangle + \langle y, z \rangle$

(ii′) $\langle x, y + z \rangle = \langle x, y \rangle + \langle x, z \rangle$

(iii) $\langle ax, y \rangle = a\langle x, y \rangle$

(iii′) $\langle x, ay \rangle = \langle x, y \rangle\bar{a}$

(iv) $\langle x, x \rangle$ is a real number ≥ 0, and $\langle x, x \rangle = 0 \Leftrightarrow x = 0$

(v) If $\{e_1, \ldots, e_n\}$ is the standard "basis" for $\mathbf{H}^n$ (i.e., $e_1 = (1, 0, \ldots, 0)$, etc.), then $\langle e_i, e_j \rangle = \delta_{ij}$.

(vi) $\langle\, , \rangle$ is nondegenerate.

(10) For $x, y \in \mathbf{H}^n$ and $A \in M_n(\mathbf{H})$, show that

$$\langle xA, y \rangle = \langle x, y\,{}^t\!\bar{A} \rangle .$$

(11) The matrix $A \in M_n(\mathbf{H})$ is said to be *symplectic* if

$$\langle xA, yA \rangle = \langle x, y \rangle$$

for all $x, y \in \mathbf{H}^n$. Show that $\mathrm{Sp}(n) = \{A \in M_n(\mathbf{H}) | A \text{ is symplectic}\}$ is a group under matrix multiplication. $\mathrm{Sp}(n)$ is called the *symplectic group*.

(12) Show that A is symplectic $\Leftrightarrow$

$$\langle Ax, Ax \rangle = \langle x, x \rangle$$

for all $x \in \mathbf{H}$.

C. Alternative and Division Algebras

Now we show that normed algebras are rather special. First they satisfy a weak kind of associativity called "alternative," and second they are division algebras.

Definition An algebra A is *alternative* if the associator $[x, y, z] = (xy)z - x(yz)$ is zero whenever two of $\{x, y, z\}$ are equal.

Proposition 6 *Any normed algebra A is alternative.*

Proof: We need to show that $[x, w, w] = 0$, $[w, x, w] = 0$ and $[w, w, x] = 0$. First we prove that for any x, w in A, we have

$$[x, w, \bar{w}] = 0\,. \tag{14}$$

To see this, note that

$$\begin{aligned} [x, w, \bar{w}] &= (xw)\bar{w} - x(w\bar{w}) \\ &= (xw)\bar{w} - x\|w\|^2 \end{aligned}$$

by Proposition 5. For any $y \in A$, note that

$$\begin{aligned} \langle (xw)\bar{w}, y\rangle &= \langle xw, yw\rangle && \text{by Proposition 3} \\ &= \langle x, y\rangle \|w\|^2 && \text{by Proposition 2} \\ &= \langle x\|w\|^2, y\rangle\,. \end{aligned}$$

But $\langle\,,\rangle$ is nondegenerate, so $(xw)\bar{w} = x\|w\|^2$ proving (14).

Quite similarly, one proves

$$[w, \bar{w}, x] = 0. \tag{15}$$

Now write $w = c + \delta$ with $c \in \mathbf{R}$ and $\delta \in \mathbf{R}^\perp$. Then

$$[x, w, w] = [x, c + \delta, c + \delta] = [x, c, c] + [x, c, \delta] + [x, \delta, c] + [x, \delta, \delta]$$

since $[\,,\,,\,]$ is trilinear. The first three terms contain $c \in \mathbf{R}$ and thus they are zero by Proposition 1; thus

$$[x, w, w] = 0 \Leftrightarrow [x, \delta, \delta] = 0\,.$$

But $\bar{\delta} = -\delta$, so that $[x, \delta, \delta] = -[x, \delta, \bar{\delta}]$ which is zero by (14). Similarly, (15) implies that $[w, w, x] = 0$.

It remains to prove that $[w, x, w] = 0$. Since $[w, w, x] = 0$, we know that

$$0 = [x + w, x + w, w]$$

and the right hand side is

$$[x, x, w] + [x, w, w] + [w, w, w] + [w, x, w] .$$

But the first three terms are zero by (14) and (15). This proves $[w, x, w] = 0$ and completes the proof of Proposition 6. ∎

Definition An algebra A with unit element is a *division algebra* if each nonzero $a \in A$ has a unique left and right inverse (i.e., there is a unique $a^{-1} \in A$ such that $aa^{-1} = a^{-1}a = 1$).

Proposition 7 *Any normed algebra A is a division algebra. In fact, given $a, b \in A$ with $a \neq 0$, the equations $ax = b$ and $xa = b$ can be solved uniquely for x.*

Proof: For $a \neq 0$, we have $a^{-1} = \frac{\bar{a}}{\|a\|^2}$ since $a\bar{a} = \|a\|^2$ (by Proposition 5).

Now note that $(b\bar{a})a = b(\bar{a}a)$ by (14), so the solution of $xa = b$ is $x = \frac{b\bar{a}}{\|a\|^2}$. Similarly, the solution of $ax = b$ is $x = \frac{\bar{a}b}{\|a\|^2}$. ∎

We conclude this section with some important formulas for a general normed algebra A. (Proofs are outlined in the exercises.)

Suppose $x, y \in A$ are orthogonal, i.e., $\langle x, y \rangle = 0$. Then

$$x\bar{y} = -y\bar{x} , \tag{16}$$

$$x(\bar{y}w) = -y(\bar{x}w) , \quad \text{and} \tag{17}$$

$$(w\bar{y})x = -(w\bar{x})y \tag{18}$$

for an arbitrary $w \in A$.

Exercises

(1) Using the fact that $2\langle x, y \rangle = x\bar{y} + y\bar{x}$ in a normed algebra A, show that

$$2\langle x, y \rangle w = [x, \bar{y}, w] + [y, \bar{x}, w] + x(\bar{y}w) + y(\bar{x}w)$$

for any $x, y, w \in A$.

(2) Show that the formula in exercise (1) reduces to

$$2\langle x, y \rangle w = x(\bar{y}w) + y(\bar{x}w)$$

by considering the associator $[x + y, \overline{x + y}, w]$.

(3) Now show that

$$2\langle x, y \rangle w = (w\bar{y})x + (w\bar{x})y$$

for all $x, y, w \in A$.

(4) Assuming that x and y are orthogonal, prove formulas (16), (17), and (18).

(5) If the (not necessarily normed) algebra A is alternative, show that $[w, \bar{w}, x] = 0$ for any $w, x \in A$. Similarly, $[x, w, \bar{w}] = 0$ and $[w, x, \bar{w}] = 0$.

D. Cayley-Dickson Process, Hurwitz Theorem

Let B be a normed algebra and suppose A is a subalgebra with $1 \in A$. (Note that A is also a normed algebra.) Then $\mathbf{R} \subseteq A$ and so $A^{\perp} \subseteq \mathbf{R}^{\perp}$. Let e be a unit vector in $A^{\perp}$. Then $e \in \mathbf{R}^{\perp}$ so that $e^2 = -1$.

Proposition 8 *Ae is orthogonal to A; i.e., $Ae \subseteq A^{\perp}$.*

Proof: Since A is a normed algebra, we know that

$$x \in A \Leftrightarrow \bar{x} \in A .$$

Now any element of Ae is of the form ae for some $a \in A$. Thus for any $b \in A$, we have

$$\langle ae, b\rangle = \langle e, \bar{a}b\rangle = 0$$

since $\bar{a}b \in A$. Hence $ae \in A^{\perp}$. ∎

Note that Ae is a subspace of the vector space B, so we can form the sum

$$A + Ae .$$

By the previous proposition, $A \cap Ae = \{0\}$ so this sum is direct. That is,

$$A \oplus Ae$$

is a subspace of B. In fact, we can make $A \oplus Ae$ into a subalgebra.

Theorem 1 *$A \oplus Ae$ is a subalgebra of B and the multiplication in $A \oplus Ae$ is given by*

$$(a + be)(c + de) = (ac - \bar{d}b) + (da + b\bar{c})e \tag{19}$$

Proof: Clearly (19) implies that $A \oplus Ae$ is a subalgebra, so we just need to prove (19). Now

$$(a + be)(c + de) = ac + (be)(de) + a(de) + (be)c ,$$

and the first term ac is already in the desired form. By Proposition 8, we have $\langle b, e\rangle = 0$, $\langle d, e\rangle = 0$, $\langle d, be\rangle = 0$, etc. So we can use the formulas (16), (17), (18). Recall (17) says that for $\langle x, y\rangle = 0$

$$x(\bar{y}w) = -y(\bar{x}w).$$

Let $x = be$ and $\bar{y} = d$ and $w = e$. Then

$$(be)(de) = -\bar{d}((\overline{be})e) = -\bar{d}((-be)e) = -\bar{d}(b(ee)) = -\bar{d}b$$

where the second equality follows from the fact that $be \in Ae \subseteq \mathbf{R}^{\perp}$, and the third is true because B is alternative.

Next, by (16) and (17), we have

$$\begin{aligned} a(de) = a(-\overline{de}) &= a(-\bar{e}\bar{d}) = a(e\bar{d}) = -\bar{e}(\bar{a}\bar{d}) \\ &= e(\overline{da}) = -(da)\bar{e} = (da)e\,; \end{aligned}$$

and finally, by (18),

$$(be)c = -(b\bar{c})\bar{e} = (b\bar{c})e\,. \quad \blacksquare$$

Exercise. Show that the conjugation on $A \oplus Ae$ is given by

$$\overline{a + be} = \bar{a} - be\,.$$

We now use Theorem 1 to motivate the Cayley-Dickson Process. Let A be a finite-dimensional real algebra with unit element and a conjugation such that $\bar{\bar{x}} = x$ and $\overline{xy} = \bar{y}\bar{x}$. We are not assuming that A is normed, but we can define the real and imaginary parts of $x \in A$ by

$$\operatorname{Re}(x) = \frac{1}{2}(x + \bar{x}) \quad \text{and} \quad \operatorname{Im}(x) = \frac{1}{2}(x - \bar{x})\,.$$

(See exercise (1).)

We make the (external) direct sum $B = A \oplus A$ into an algebra with unit element by defining

$$(a, b)(c, d) = (ac - \bar{d}b, da + b\bar{c})\,;$$

and we define conjugation on B by

$$\overline{(a, b)} = (\bar{a}, -b)\,.$$

Exercise. Show that this conjugation on B satisfies $\bar{\bar{x}} = x$ and $\overline{xy} = \bar{y}\bar{x}$. (Thus we can repeat this construction, forming $B \oplus B$, etc.) What are the real and imaginary parts of $(a, b) \in B$?

We say that B is obtained from A by the *Cayley-Dickson Process.*

So far, we have $\mathbf{C} = \mathbf{R} \oplus \mathbf{R}$, $\mathbf{H} = \mathbf{C} \oplus \mathbf{C}$, and $\mathbf{O} = \mathbf{H} \oplus \mathbf{H}$ by using the Cayley-Dickson Process (see (2′) and the exercises of section B).

Now we regard A as a subalgebra of $B = A \oplus A$ by including it as the first factor. If we choose a unit vector e from the second factor, then we can write B as the internal direct sum

$$B = A \oplus Ae$$

with multiplication given by (19).

Theorem 2 *Suppose $B = A \oplus Ae$ is obtained from A by the Cayley-Dickson Process. Then*

(i) *B is commutative $\Leftrightarrow A = \mathbf{R}$,*

(ii) *B is associative $\Leftrightarrow A$ is associative and commutative, and*

(iii) *B is alternative $\Leftrightarrow A$ is associative.*

Proof: Suppose $x = a + \alpha e$, $y = b + \beta e$, and $z = c + \gamma e$ are elements of B with $a, \alpha, b, \beta, c, \gamma \in A$.

We will state some rather incredible-looking formulas, show that these prove the theorem, and then show that they are true by using the Cayley-Dickson formula (19).

$$
\begin{aligned}
(20)\quad [x,y] &= [a,b] + 2\,\mathrm{Im}\,(\bar{\alpha}\beta) + 2(\beta\,\mathrm{Im}\,(a) - \alpha\,\mathrm{Im}\,(b))e \\
(21)\quad [x,y,z] &= [a,\bar{\gamma}\beta] + [\bar{b},\bar{\gamma}\alpha] + [c,\bar{\beta}\alpha] + \\
&\quad (\alpha\,[\bar{b},\bar{c}] + \beta\,[a,\bar{c}] + \gamma\,[a,b] + \alpha\,[\bar{\beta},\gamma] + [\alpha,\gamma]\,\bar{\beta} + \gamma\,[\alpha,\bar{\beta}]\,)e \\
(22)\quad [x,\bar{x},y] &= [a,\bar{\beta},\alpha] + [\alpha,\bar{b},a]e
\end{aligned}
$$

(i) $\Leftarrow$

Suppose $A = \mathbf{R}$ so that all of the terms in the right hand side of (20) zero. Thus B is commutative.

(i) $\Rightarrow$ (by contrapositive)

Suppose $A \neq \mathbf{R}$ so that $\dim A \geq 2$. Now $\mathbf{R} = \mathrm{Span}(1) \subsetneq A$ so that A contains $\mathbf{C} = \mathbf{R} \oplus \mathbf{R}$ by Theorem 1. But then B must contain $\mathbf{H} = \mathbf{C} \oplus \mathbf{C}$ so that B is not commutative.

(ii) $\Leftarrow$

If A is commutative, then any commutator with both entries from A will be zero. This is precisely the case in the right hand side of (21), so B must be associative.

(ii) $\Rightarrow$

If B is associative, then A certainly is; furthermore, the associator $[x,y,z]$ is always zero. In particular, the A-part in (21)

$$[a,\bar{\gamma}\beta] + [\bar{b},\bar{\gamma}\alpha] + [c,\bar{\beta}\alpha]$$

is always zero. Let $a, \beta \in A$ be arbitrary, let $\gamma = 1$, and let $b = c = 0$. Then we see that $[a,\beta] = 0$ so A is commutative.

(iii) $\Leftarrow$

If A is associative, then the right hand side of (22) vanishes. Thus

$$[x,\bar{x},y] = 0$$

for all $x, y \in B$. Looking at the proof of Proposition 6, we see that this is sufficient to prove that B is alternative.

(iii) $\Rightarrow$

If B is alternative, then $[x, \bar{x}, y] = 0$ (see exercise (5) of the previous section). By (22), we have $[a, \bar{\beta}, \alpha] = 0$ for any $a, \bar{\beta}, \alpha \in A$ so A is associative.

Now we show that the formulas (20), (21), and (22) hold.

Using (19), we calculate

$$xy = (a + \alpha e)(b + \beta e) = (ab - \bar{\beta}\alpha) + (\beta a + \alpha\bar{b})e$$

and

$$yx = (b + \beta e)(a + \alpha e) = (ba - \bar{\alpha}\beta) + (\alpha b + \beta\bar{a})e\,.$$

Thus

$$xy - yx = ab - ba + (\bar{\alpha}\beta - \bar{\beta}\alpha) + (\beta(a - \bar{a}) - \alpha(b - \bar{b}))e\,.$$

Recall that $2\,\mathrm{Im}\,(w) = w - \bar{w}$, so this last equation becomes (20).

To establish (21), we must assume that A is associative. (This is no restriction since A is always associative in case (ii).)

Now $[x, y, z] = (xy)z - x(yz)$ and using (19), we calculate

$$\begin{aligned}(xy)z &= \{(ab - \bar{\beta}\alpha) + (\beta a + \alpha\bar{b})e\}(c + \gamma e)\\ &= (ab - \bar{\beta}\alpha)c - \bar{\gamma}(\beta a + \alpha\bar{b})\\ &\quad + \{\gamma(ab - \bar{\beta}\alpha) + (\beta a + \alpha\bar{b})\bar{c}\}e\end{aligned}$$

and

$$\begin{aligned}x(yz) &= a(bc - \bar{\gamma}\beta) - (\bar{b}\bar{\gamma} + c\bar{\beta})\alpha\\ &\quad + \{(\gamma b + \beta\bar{c})a + \alpha(\bar{c}\bar{b} - \bar{\beta}\gamma)\}e\,.\end{aligned}$$

So we have

$$[x, y, z] = \left\{\begin{array}{l}(ab)c - (\bar{\beta}\alpha)c - \bar{\gamma}(\beta a) - \bar{\gamma}(\alpha\bar{b})\\ -\,a(bc) + a(\bar{\gamma}\beta) + (\bar{b}\bar{\gamma})\alpha + (c\bar{\beta})\gamma\end{array}\right\} + \left\{\begin{array}{l}\gamma(ab) - \gamma(\bar{\beta}\alpha) + (\beta a)\bar{c} + (\alpha\bar{b})\bar{c}\\ -(\gamma b)a - (\beta\bar{c})a - \alpha(\bar{c}\bar{b}) + \alpha(\bar{\beta}\gamma)\end{array}\right\}e\,. \tag{23}$$

Since A is associative, $(ab)c - a(bc) = 0$ in the A-part. Also, in the A-part, we have

$$\begin{aligned}c(\bar{\beta}\alpha) - (\bar{\beta}\alpha)c &= [c, \bar{\beta}\alpha] &&\text{(terms 8 and 2)}\\ a(\bar{\gamma}\beta) - (\bar{\gamma}\beta)a &= [a, \bar{\gamma}\beta] &&\text{(terms 6 and 3)}\\ \bar{b}(\bar{\gamma}\alpha) - (\bar{\gamma}\alpha)\bar{b} &= [\bar{b}, \bar{\gamma}\alpha] &&\text{(terms 7 and 4)}\,.\end{aligned}$$

Hence the A-part of (23) is the same as in (21).

For the Ae-part, we have

$$\begin{aligned}\gamma(ab) - \gamma(ba) &= \gamma[a, b] &&\text{(terms 1 and 5)}\\ \beta(a\bar{c}) - \beta(\bar{c}a) &= \beta[a, \bar{c}] &&\text{(terms 3 and 6)}\\ \alpha(\bar{b}\bar{c}) - \alpha(\bar{c}\bar{b}) &= \alpha[\bar{b}, \bar{c}] &&\text{(terms 4 and 7)}\,.\end{aligned}$$

We are left with the terms $\alpha(\bar{\beta}\gamma) - \gamma(\bar{\beta}\alpha)$ which we rewrite as

$$\begin{aligned}\alpha\bar{\beta}\gamma - \alpha\gamma\bar{\beta} + \alpha\gamma\bar{\beta} - \gamma\alpha\bar{\beta} + \gamma\alpha\bar{\beta} - \gamma\bar{\beta}\alpha \\ = \alpha\,[\bar{\beta},\gamma] + [\alpha,\gamma]\,\bar{\beta} + \gamma\,[\alpha,\bar{\beta}]\,,\end{aligned}$$

so (21) holds.

To establish (22), we must assume that A is alternative (which is the case in (iii)). We calculate

$$x\bar{x} = (a + \alpha e)(\bar{a} - \alpha e) = a\bar{a} + \bar{\alpha}\alpha$$

and so

$$(x\bar{x})y = (a\bar{a} + \bar{\alpha}\alpha)b + (a\bar{a} + \bar{\alpha}\alpha)\beta e\,. \tag{24}$$

Also,

$$\bar{x}y = (\bar{a} - \alpha e)(b + \beta e) = (\bar{a}b + \bar{\beta}\alpha) + (\beta\bar{a} - \alpha\bar{b})e$$

so that

$$\begin{aligned} x(\bar{x}y) &= (a + \alpha e)\{(\bar{a}b + \bar{\beta}\alpha) + (\beta\bar{a} - \alpha\bar{b})e\} \\ &= a(\bar{a}b + \bar{\beta}\alpha) - (a\bar{\beta} - b\bar{\alpha})\alpha \\ &\quad + \{(\beta\bar{a} - \alpha\bar{b})a + \alpha(\bar{b}a + \bar{a}\beta)\}e \\ &= a(\bar{a}b) + a(\bar{\beta}\alpha) - (a\bar{\beta})\alpha + (b\bar{\alpha})\alpha \\ &\quad + \{(\beta\bar{a})a - (\alpha\bar{b})a + \alpha(\bar{b}a) + \alpha(\bar{a}\beta)\}e \\ &= (a\bar{a})b + b(\bar{\alpha}\alpha) - [a,\bar{\beta},\alpha] \\ &\quad + \{\beta(\bar{a}a) + (\alpha\bar{\alpha})\beta - [\alpha,\bar{b},a]\,\}e \end{aligned}$$

where the last equality follows from the fact that A is alternative.

Now note that $w\bar{w} \in \mathbf{R}$ for any $w \in A$. (<u>Exercise</u>. Show this.) So we have

$$\begin{aligned} x(\bar{x}y) &= (a\bar{a} + \bar{\alpha}\alpha)b - [a,\bar{\beta},\alpha] \\ &\quad + \{(a\bar{a} + \bar{\alpha}\alpha)\beta - [\alpha,\bar{b},a]\,\}e\,. \end{aligned} \tag{25}$$

Subtracting (25) from (24) gives (22) which finishes the proof of Theorem 2. ∎

An important corollary of Theorem 2 is

Theorem 3 (Hurwitz Theorem) *The only real normed algebras are* $\mathbf{R}, \mathbf{C}$, $\mathbf{H}$, *and* $\mathbf{O}$.

Proof: Let B be a normed algebra and let $\mathbf{R} = \mathrm{Span}(1)$. If $B = \mathbf{R}$, then we are done. If not, then $\mathbf{R}^{\perp} \neq \{0\}$ so we can choose a unit vector $i \in \mathbf{R}^{\perp}$. By Theorem 1, $\mathbf{C} = \mathbf{R} \oplus \mathbf{R}i$ is a subalgebra of B.

If $B = \mathbf{C}$, then we are done; otherwise, we choose a unit vector $j \in \mathbf{C}^{\perp}$. Then $\mathbf{H} = \mathbf{C} \oplus \mathbf{C}j$ is a subalgebra of B.

If $B = \mathbf{H}$, we are done. If not, we choose a unit vector $e \in \mathbf{H}^{\perp}$ so that $\mathbf{O} = \mathbf{H} \oplus \mathbf{H}e$ is a subalgebra of B.

If $\mathbf{O}$ is not all of B, then B contains the subalgebra $\mathbf{O} \oplus \mathbf{O}\tau$ where $\tau \in \mathbf{O}^{\perp}$ is a unit vector. But B is a normed algebra so that $\mathbf{O} \oplus \mathbf{O}\tau$ is also normed. Hence $\mathbf{O} \oplus \mathbf{O}\tau$ is alternative (by Proposition 6) so that $\mathbf{O}$ must be associative by Theorem 2. Since $\mathbf{O}$ is not associative, we must have $B = \mathbf{O}$. ∎

Exercises

(1) Let A be a finite-dimensional real algebra with unit element, and choose a basis $\{v_1, \ldots, v_n\}$ for A such that $v_1 = 1$. For $x = x_1v_1 + \cdots + x_nv_n \in A$, define

$$\bar{x} = x_1 - x_2v_2 - \cdots - x_nv_n$$

and show that this conjugation satisfies $\bar{\bar{x}} = x$ and $\overline{xy} = \bar{y}\bar{x}$. Also, show that $\mathrm{Re}(x) = \frac{1}{2}(x + \bar{x})$ is, indeed, in $\mathrm{Span}(1) = \mathbf{R}$.

(2) Show that the Cayley-Dickson process works. That is, show that the multiplication defined on $B = A \oplus A$ makes B into an algebra with unit element.

(3) Consider the algebra $B = \mathbf{O} \oplus \mathbf{O}$ obtained via the Cayley-Dickson process. Show that

$$\|x\|^2 = x\bar{x}$$

still holds, where $\| \ \|$ is the standard norm on $\mathbf{R}^{16}$. Conclude that B is a division algebra. However, given $a, b \in B$, the equation

$$ax = b$$

cannot always be solved uniquely for x since B is not alternative.

(4) Let A be the subalgebra (with unit element) which is generated by any two elements of $\mathbf{O}$. Show that A is associative. (Hint: If $x, y \in \mathbf{O}$ generate A, write these as $x = x_1 + x_2, y = y_1 + y_2$ where $x_1 = \mathrm{Re}(x)$ and $x_2 = \mathrm{Im}\,(x)$, etc. If $A \neq \mathbf{R}$, consider $\mathbf{R} \oplus \mathbf{R}\epsilon \subseteq A$ where $\epsilon = \frac{x_2}{\|x_2\|}$.)

(5) We have shown that $x\bar{x} = \|x\|^2$ and $\overline{xy} = \bar{y}\bar{x}$ hold in a general normed algebra. Prove these directly for $\mathbf{O}$ (without assuming that $\mathbf{O}$ is normed).

(6) Use exercises (3) and (4) to give a proof that $\mathbf{O}$ is a normed algebra (i.e., satisfies (1)). This also shows that $\mathbf{R}, \mathbf{C}$, and $\mathbf{H}$ are normed.

(7) Given $x, y, z \in \mathbf{O}$, prove the identities

$$\begin{aligned} (xyx)z &= x(y(xz))\,, \\ z(xyx) &= ((zx)y)x\,, \quad \text{and} \\ (xy)(zx) &= x(yz)x\,. \end{aligned}$$

Note that the expression xyx makes sense since $\mathbf{O}$ is alternative. (Hint: If any two of $\{x, y, z\}$ are equal, then these equations reduce to zero by exercise (3). Show that these formulas are linear in y and z, so we can assume x, y, and z are orthogonal. Now apply exercises (2) and (3) from the previous section.)

Bibliography

[1] Morton L. Curtis, *Matrix Groups*, second edition, Universitext Series, Springer-Verlag New York, Inc. (1984).

[2] Paul R. Halmos, *Finite-Dimensional Vector Spaces*, second edition, Undegraduate Texts in Mathematics Series, Springer-Verlag New York, Inc. (1974).

[3] F. Reese Harvey and H. Blaine Lawson, Jr., "Calibrated Geometries," *Acta. Math.* 148 (1982), 47-157.

[4] I.N. Herstein, *Topics in Algebra*, second edition, John Wiley & Sons (1975).

[5] Serge Lange, *Linear Algebra*, second edition, Addison -Wesley Publishing Co., Inc. (1971).

[6] Seymour Lipschutz, *Linear Algebra*, Schaum's Outline Series, McGraw-Hill Book Co. (1968).

Index